T0134707

Springer Theses

Recognizing Outstanding Ph.D. Research

Aims and Scope

The series "Springer Theses" brings together a selection of the very best Ph.D. theses from around the world and across the physical sciences. Nominated and endorsed by two recognized specialists, each published volume has been selected for its scientific excellence and the high impact of its contents for the pertinent field of research. For greater accessibility to non-specialists, the published versions include an extended introduction, as well as a foreword by the student's supervisor explaining the special relevance of the work for the field. As a whole, the series will provide a valuable resource both for newcomers to the research fields described, and for other scientists seeking detailed background information on special questions. Finally, it provides an accredited documentation of the valuable contributions made by today's younger generation of scientists.

Theses are accepted into the series by invited nomination only and must fulfill all of the following criteria

- They must be written in good English.
- The topic should fall within the confines of Chemistry, Physics, Earth Sciences, Engineering and related interdisciplinary fields such as Materials, Nanoscience, Chemical Engineering, Complex Systems and Biophysics.
- The work reported in the thesis must represent a significant scientific advance.
- If the thesis includes previously published material, permission to reproduce this must be gained from the respective copyright holder.
- They must have been examined and passed during the 12 months prior to nomination.
- Each thesis should include a foreword by the supervisor outlining the significance of its content.
- The theses should have a clearly defined structure including an introduction accessible to scientists not expert in that particular field.

More information about this series at http://www.springer.com/series/8790

Gauthier Guillaume

Transport and Turbulence in Quasi-Uniform and Versatile Bose-Einstein Condensates

Doctoral Thesis accepted by
The University of Queensland, QLD, Australia

Springer

Author
Dr. Gauthier Guillaume
School of Mathematics and Physics
The University of Queensland
Brisbane, QLD, Australia

Supervisor
Dr. Tyler Neely
School of Mathematics and Physics
The University of Queensland
Brisbane, QLD, Australia

ISSN 2190-5053 ISSN 2190-5061 (electronic)
Springer Theses
ISBN 978-3-030-54969-5 ISBN 978-3-030-54967-1 (eBook)
https://doi.org/10.1007/978-3-030-54967-1

© The Editor(s) (if applicable) and The Author(s), under exclusive license to Springer Nature Switzerland AG 2020
This work is subject to copyright. All rights are solely and exclusively licensed by the Publisher, whether the whole or part of the material is concerned, specifically the rights of translation, reprinting, reuse of illustrations, recitation, broadcasting, reproduction on microfilms or in any other physical way, and transmission or information storage and retrieval, electronic adaptation, computer software, or by similar or dissimilar methodology now known or hereafter developed.
The use of general descriptive names, registered names, trademarks, service marks, etc. in this publication does not imply, even in the absence of a specific statement, that such names are exempt from the relevant protective laws and regulations and therefore free for general use.
The publisher, the authors and the editors are safe to assume that the advice and information in this book are believed to be true and accurate at the date of publication. Neither the publisher nor the authors or the editors give a warranty, expressed or implied, with respect to the material contained herein or for any errors or omissions that may have been made. The publisher remains neutral with regard to jurisdictional claims in published maps and institutional affiliations.

This Springer imprint is published by the registered company Springer Nature Switzerland AG
The registered company address is: Gewerbestrasse 11, 6330 Cham, Switzerland

Dédié à mes parents pour leur amour et leur soutien infaillible au cours des années

Supervisor's Foreword

Ultracold atoms and Bose-Einstein Condensates (BECs) are nearly ideal experimental systems for investigating a wide variety of condensed matter physics. Some of these aspects arise from the dilute nature of these systems, such as the simplicity of ultracold collisions, for example, but they also arise from the wide variety of refined experimental techniques available to the researcher. Among the more recent advanced techniques, the microscopic control of the BEC density and phase using optical fields has enabled much of the recent experimental progress in the field. Of particular relevance to this thesis, advanced optical control has yielded new experiments in quantum turbulence and new developments in the application-oriented field of atomtronics.

Dr. Gauthier's thesis foremost develops and demonstrates the application of directly imaged Digital Micromirror Devices (DMDs) for trapping and configuring BECs. After first demonstrating the high-resolution DMD pattern projection, trapping, and imaging of BECs, Dr. Gauthier then develops a feed-forward technique permitting engineering of both the density and phase of the BEC. These techniques establish greater control of BECs that will interest readers who seek to enable future cold atoms experiments with finely structured light. Dr. Gauthier applies these optical trapping techniques to investigate two emerging areas of focus in superfluid physics: atomtronic oscillators and two-dimensional (2D) quantum turbulence.

Atomtronics has emerged as a framework for the building of atom devices analogous to their electrical counterparts, but where the flow of neutral atoms takes the place of charge carriers. Dr. Gauthier's thesis demonstrates that for lumped atomtronics circuits, such as the dumbbell inductor-capacitor (LC) oscillator investigated, an approach based on acoustics rather than direct electronic analogues can more accurately predict the behaviour of these superfluid circuits. These insights will be important to readers who seek approaches for building superfluid circuit devices, such as inertial sensors.

Among Dr. Gauthier's most substantial results are the observation, 70 years after its prediction, of the vortex cluster states of 2D turbulence that were predicted by Lars Onsager in 1949. By establishing a uniform quasi-2D condensate, and dynamically controlling a DMD to inject vortices, Dr. Gauthier shows that vortex clusters can persist in 2D superfluids for many seconds. These results demonstrate that point-vortex equilibria can be observed in superfluids, and such states are robust even when the entire BEC system is far from equilibrium.

In summary, Dr. Gauthier's thesis will be of interest to a broad range of the Springer Thesis audience. It will be of interest to those seeking a technical manual for implementing DMDs and high-resolution optical trapping and imaging of BECs, or those readers seeking to further develop atomtronic technology and sensors based on trapped superfluids. It will be of further interest for those readers interested in the progress and enabling techniques for the 2D quantum turbulence research that this thesis pioneers.

Brisbane, Australia
June 2020

Dr. Tyler Neely

Abstract

Turbulence, motion characterized by chaotic changes in pressure and flow velocity, is a challenging problem in physics. However, its underlying properties are found to be universal and do not depend on the host fluid. Meanwhile, transport phenomena, irreversible exchange processes due to the statistical continuous random motion of particles, although being complicated from a microscopic point of view, can often be modelled quite simply by tracking macroscopic quantities of interest in the system. In this thesis, using atomic Bose-Einstein condensates, we study both these phenomena inside a quantum fluid using a highly configurable BEC platform developed to provide arbitrary dynamic control over the 2D superfluid system. Furthermore, the experiments are modelled using the Gross-Pitaevskii equation, the point vortex model and the hydrodynamic equations.

After theoretical background and introduction to the apparatus, the technique of direct imaging of a digital micromirror device is described, which achieves the highly versatile and dynamic 2D potentials that facilitate the experimental studies described. Superfluid transport through a mesoscopic channel of tuneable length and width is next described. By investigating low amplitude oscillations and their dependence on the system parameters, a resistor, capacitor, and inductor model is used to model the transport. Surprisingly, the "contact inductance" of the channel at the reservoirs is a dominant effect for a significant portion of the parameter range. The resistive transport for high initial bias is also studied, where we find turbulent and sound dissipation regimes of energy dissipation during the transport. Next, the transport between two reservoirs initially prepared at different temperatures, but with similar particle number, was explored.

Our 2D superfluid system, with hard-wall confinement, provides an ideal experimental system for the study of 2D quantum turbulence. The system is utilized to demonstrate the first experimental realization of large Onsager vortex clusters in the negative absolute temperature regime, through the injection of high energy clusters into the 2D superfluid. The clusters are found to be surprisingly stable for long time periods. The vortex cluster energy loss rate is studied while changing the system parameters, suggesting thermal damping is the dominant loss mechanism.

The techniques and results presented in this thesis open up new avenues for the study of quantum fluids, be it by providing a concise atomtronic model for predicting superfluid transport or expanding the accessible parameters space available to fundamental studies of turbulence. The realization of negative temperature vortex distributions, long ago predicted by Onsager, open up the experimental study of the full phase-diagram of 2D vortex matter. The refinement of optical trapping techniques for BECs presents new and promising directions for future BEC experiments in configured potentials.

Publications Related to this Thesis

1. **G. Gauthier**, I. Lenton, N. McKay Parry, M. Baker, M. Davis, H. Rubinsztein-Dunlop and T. W. Neely, Direct imaging of a digital-micromirror device for configurable microscopic optical potentials, *Optica*, **3**, 1136–1143 (2016).
2. **G. Gauthier***, M. T. Reeves*, X. Yu, A. S. Bradley, M. Baker, T. A. Bell, H. Rubinsztein-Dunlop, M. J. Davis, T. W. Neely, Giant vortex clusters in a two-dimensional quantum fluid, *Science*, **364**, 1264–1266 (2019).
3. **G. Gauthier**, Stuart S. Szigeti, Matthew T. Reeves, Mark Baker, Thomas A. Bell, Halina Rubinsztein-Dunlop, Matthew J. Davis, and Tyler W. Neely, Quantitative Acoustic Models for Superfluid Circuits, *Physical Review Letters*, **123**, 260402 (2019).

Acknowledgements

This thesis is the culmination of many hours of work, to which a vast number of people contributed some of whom may not even know it. Without their insights and encouragement, this document would probably never have seen the light of day.

First and foremost, I would like to thank my primary supervisor, Dr. Tyler Neely, for his endless encouragement, mentorship, humour, and ongoing emotional support over the past 4 years. Not only did you always manage to make time for me when I needed it, but your endless enthusiasm, passion for science and creativity served as an inspiration throughout my stay at UQ. You always made me feel like an important part of the team, and I greatly appreciate your friendship. I'm also very grateful to my secondary supervisors, Dr. Mark Baker and Prof. Halina Rubinsztein-Dunlop, who have given me feedback on all my work over the year and were able to provide valuable insights. All of you were patient when patience was required, demanding when needed, but always encouraging, even when I accidentally burnt the Prosilica camera. For having had such support I am grateful.

I learnt a lot from Prof. Matthew Davis and Dr. Matthew Reeves and would like to thank you both for being willing to answer my questions and patience for basic physics discussions which have been invaluable in my career as a graduate student. Thanks to Jake Glidden, Dr. Stuart Szigeti, and Dr. Michael Bromley, for teaching me numerical techniques and helping me with all the basic questions I kept coming up with. Thanks also to our outside theory collaborators Dr. Ashton Bradley, and Dr. Xiaoquan Yu whose expertise and insights have led to a better understanding of our experiments leading to better joint publications than they would have been otherwise.

Throughout my Ph.D., I was fortunate to attend many conferences and schools in Australia and abroad. These trips were a highlight for me and I am grateful to everyone I met during my travels and who helped make these experiences so enjoyable and memorable. Thanks to the IONS KOALA 2017 team for making organizing a conference so enjoyable and in particular to Changqiu 'Sarah' Yu for doing such a great job as the president of our organizing committee. Special thanks to Joyce Wang for all the help with regard to managing the IONS KOALA 2017 funds and for putting up with all my unceasing inquiries about basic finance knowledge; without you I would not have been half the treasurer I was. Hosting a conference was one of the most challenging and rewarding experiences of my Ph.D.

The experiments presented in this dissertation would not have been possible without the help of those who came before me, with most of the credit going to Nicholas McKay Parry and Issac Lenton. Their preliminary work in building the apparatus, in the face of many setbacks, has allowed for the remarkable science apparatus presented here. I have also been extremely lucky to work and share the lab with talented experimentalists and all around good people. Thomas Bell's technical expertise and ways of thinking about physics often provided alternate and much appreciated solutions to my problems. Alexander Pritchard made taking data in the lab entertaining and has been a wonderful friend. Kwan Goddard Lee has been a most excellent minion over the past year and with the experiment now in his capable hands I can't wait to see where it goes.

I would like to thank the administration staff—Ruth Forrest, Murray Kane, Tara Massingham, Danette Peachey, Tara Roberson, Lisa Walker, and Angie Bird—who made administration less of a burden than it would have been otherwise, and thank you to Sam Zammit who was very helpful any time a software issue would pop-up. Thanks to the machine shop staff, Neil and David, for accommodating my last minute custom part designs that needed to be finished yesterday and for pointing out the error of my ways when I was trying to overcomplicate my mechanical designs.

Finally, I would like to thank all of those who went adventuring with me across Australia and New Zealand, for reminding me that there is a world outside of physics and the lab and preventing a descent into madness. Special mention to Severin Charpignon, Yevgeni Bondarchuk and Kristen Keidel for accompanying me on my amazing (sometimes terrifying) hiking, backpacking, diving, and kayaking adventures. I would also like to thank my non-physicist friends for helping maintain contact with the outside world. Special thanks to Mentari Widiastuti, Michelle Leung, Evan Weber, Ashley Ervin and Samuel Boudreau for providing laughter, conversation and some much-needed ranting over beers. Thanks to my landlords Kerry Gray and Gloria Gray giving me a place to thrive during my studies and for all of the great diners over the years. Last but not least, thanks to my family—Mum, Dad, Connie and Kéven—who have always loved me unconditionally and supported me in both my academic and other pursuits despite being confused about what it is I do.

This research was supported by: an Australian Government Research Training Program Scholarship; the ARC Centre for Engineered Quantum Systems (Project No. CE110001013); the ARC Centre for Engineered Quantum Systems (Project No. CE170100009); and an ARC Discovery Project (No. DP160102085).

Contents

Abbreviations and Symbols

Abbreviations

2D	Two-Dimensional
2DMOT	2D Magneto Optical Trap
2DQT	2D Quantum Turbulence
3D	Three-Dimensional
3DMOT	3D Magneto Optical Trap
μ-wave	Microwave
AC	Alternating Current
AOM	Acousto Optical Modulator
AR	Anti-Reflective
BCS	Bipolar Current Source
BEC	Bose-Einstein Condensate
CCD	Charge-Coupled Device
DC	Direct Current
DLP	Digital Light Processing
DMD	Digital Micromirror Device
DoF	Depth of Field
FFT	Fast Fourier Transform
FWHM	Full Width at Half-Maximum
HWHM	Half Width at Half-Maximum
K	Potassium
LC	Inductor-Capacitor circuit
LMS	Least Mean Square
MOT	Magneto Optical Trap
MTF	Modulation Transfer Function
MW	Microwave
OD	Optical Density
ODT	Optical Dipole Trap
PBS	Polarizing Beam Splitter

PID	Proportional-Integral-Derivative
PM	Polarization Maintaining
PSF	Point Spread Function
QT	Quantum Turbulence
Rb	Rubidium
RC	Resistor-Capacitor circuit
RF	Radio Frequency
RLC	Resistor-Inductor-Capacitor circuit
RMS	Root Mean Squared
TOF	Time of Flight
UHV	Ultra High Vacuum

Symbols

$\widehat{H}$	Hamiltonian
$\mathscr{L}$	Lagrangian

Chapter 1
Introduction

This introduction serves to outline the historical context, motivation, and general overview of the phenomena that are the main subject of this thesis, focusing on transport and turbulence in superfluids. These phenomena are introduced from their classical conception and the link with modern work in their quantum counter parts is outlined, to show that deeper understanding in one regime leads to better understanding into the other. Along the way, the dilute-gas Bose-Einstein condensate (BEC) that are used for our studies is introduced. For a more in depth mathematical foundation of these phenomena see Chap. 2.

1.1 Transport

Transport involves the movement of mass, energy, charge, angular momentum, magnetism, spin, etc. inside a system of interest, or the exchange of those same quantities between separate systems. Transport is usually studied as part of physics, engineering, and chemistry, but can be found in most out of equilibrium systems and is a still very active field of research, even for classical transport. Although the theoretical bases underlying transport phenomena were virtually complete by the 1950s [1], the infinite number of unique boundary conditions and initial conditions imply that the numerical study of transport will continue well into the future. An example of modern day research into the transport of energy is the origin of rogue waves in three-dimensional oceanic systems, which is still being investigated [2]. Theoretical models for transport in far-from-equilibrium conditions are still being developed [3] and the fundamentals of quantum transport are still being studied in novel material systems [4].

© The Editor(s) (if applicable) and The Author(s), under exclusive license to Springer Nature Switzerland AG 2020

G. Guillaume, *Transport and Turbulence in Quasi-Uniform and Versatile Bose-Einstein Condensates*, Springer Theses,
https://doi.org/10.1007/978-3-030-54967-1_1

Transport has been studied since the 1700s, with Bernoulli [5], Euler [6] and Newton [1] developing models of transport for ideal collisionless fluids. Around 1820 the physicist Claude-Louis Navier and physicist-mathematician George Gabriel Stokes independently proposed a set of equations to describe the motion of viscous fluids, now called the Navier-Stokes equations, which are still the subject of current study. Around the same time the equations of heat flux and conduction were developed by Fourier to explain the transport of kinetic energy inside different media. In a treatise published in 1827, physicist Georg Ohm similarly developed the equations to describe the movement of charges through resistive materials. Fick's first and second law of diffusion were first published in 1855 [7]. In the late 1800s, Maxwell and Boltzmann derived statistical equations to describe the distribution of momentum in mono-atomic gasses and Boltzmann went on to use statistical mechanics to describe the link between the microscopic properties of atoms/molecules (mass, charge, orbital, and vibrational modes) and the macroscopic properties of matter (viscosity, diffusion, and thermal transport). At the beginning of the 20th century, heat convection, thermal radiation, and convective mass transport would be established and studied [1]. In the mid-1900 new transport effects such as thermal diffusion, the diffusion-thermo effect, and forced diffusion were investigated by Onsager, Eckart and others. By then the fundamental equations for transport phenomena were established, but due to lack of computing power only the most simple of scenarios, most having analytical solutions, were investigated. Numerical simulations of transport properties and realistic equations were enabled with the development of computers and continue to this day [1].

Transport continues to be studied in quantum systems where the dynamics of single particles are well described by the Schrödinger equation. In a similar fashion to classical transport, interactions between multiple particles complicate the dynamics, and when enough particles are present statistical mechanics can be used to simplify the description of the behaviour of the physical system into a topography of macroscopic entities that approximate the behaviour of the underlying system. This process is known as the lumped-element model. Studies have focussed on developing similar components to electronic circuits which include transistors [8, 9], batteries [10] and mesoscopic oscillators [11–13]. To get a better grasp of the underlying physics Josephson oscillators [14–17] and quantised conductance channels [18, 19] are actively being studied. The hysteresis of quantised circulations [20, 21] around toroidal waveguides has also been studied for metrological applications [22, 23]. Beyond replicating classical components using phase coherent matter, these studies are useful in the development of the next generation of high-precision sensor such as gravimeters [24, 25] and gravity gradiometers [26] and their related protocols [27–31] which will be useful for gravitational astronomy [32, 33]. Many Sagnac interferometric protocols have also been proposed to make precision measurements using quantum systems [34–43], and some have been realized for magnetic gradiometry [44] and inertial rotation [45–47].

In this thesis transport in quantum systems is discussed in Chap. 5, where we investigate the transport of superfluid between reservoirs and show that the transport near equilibrium, equivalent to laminar flow in classical systems, is well approximated by classical acoustic model. Far from equilibrium, we experimentally demonstrate that turbulence and sound excitations form the basis for resistive transport and investigate the regimes in which these different mechanisms are present.

1.2 Turbulence

As described above, fluid transport can act to generate turbulence. Turbulence is also a subject of study with a long history, due to its prevalence across a wide range of physical systems. We commonly experience turbulence in flight, hear it in the waves crashing at a beach, observe it in the color patterns that appear on the surface of soap bubbles [48], take advantage of it when stirring a coffee cup, smell it in the mixing of smoke in air from a camp-fire and feel it in flow of blood through our heart cavity after strenuous exercise. The phenomena is very general and encompasses the departure from smooth flow in air, water, or other types of fluid. This departure is characterized by irregular motion of the fluid and it usually manifests in systems where the transport of energy, matter, magnetism, momentum, etc. has non-linear properties [49–54]. Examples of fields where turbulence plays a role include: the propagation of light through non-linear optics [55–58], the flow of water, in magnetohydrodynamics [59–63] (a.k.a. magneto-fluid dynamics or hydromagnetics), and in the formation of galaxies. Turbulence is not only important to a wide range of physical systems both natural and man-made, but it is found throughout physics at a wide range of length scales from nanometres to galactic length and was studied as far back as Leonardo da Vinci [64]. Turbulence plays an important role in the dynamics of neutron star crust [65], atmospheric patterns [66–68] such as the big red spot on Jupiter [69] and the transport of pollutants in the atmosphere, the distribution of phytoplankton in oceans [70–72], the locomotion of water strider [73], and the frequency splitting of sound modes in superfluid helium [74, 75]. Not only is turbulence scale invariant, it is also chaotic, in both time and space [76], and statistically predictable which makes it a subject of interest from a purely mathematical standpoint [77]. However, turbulence is so complicated that even after centuries of research it remains unsolved. Even the simplest scenario imaginable, fast flowing water in a pipe [78], has no known solution and it is unknown if one even exists. It is one of seven unsolved million-dollar "Millennium Prize" problems set by the Clay Mathematics Institute and is widely recognised as an important unsolved problem of classical physics.

Turbulence has major economic impact on society due to the heat, and the drag it generates. These show up as impacts through loss of revenue during industrial production and increase CO_2 emission during transport. In countless engineering applications turbulence is viewed as harmful such as when designing the aerodynamical properties of vehicles, designing of swim suits for athlete swimmers, and in transport through pipelines [79]. In all three of these applications, turbulence leads to increased energy loss through heating of the environment due to deviation of the flow from a laminar flow. Turbulence can also cause unwanted mechanical vibrations that can be a detriment in the average lifespan of an application such as a plane or a pipeline. Not all turbulence is detrimental, it can actually be harnessed since it facilitates the rapid transport of energy, momentum and mass throughout the fluid. Turbulence can be exploited to aid mixing [80] or heat transfer in industrial processes such as in combustion engines [81, 82], in aerodynamic spoilers which control lift on aircraft, in the design of dimples on the golf ball and rifling of gun chambers which are used to extend and stabilize the flight paths in their respective applications. Turbulence will also play a role in future quantum technologies such as room temperature superconductors [83], topological quantum computers [84] and extremely precise sensors [74].

Despite several centuries of investigation, a strict mathematical definition of turbulence eludes fluid researchers. Some texts even refuse to give a definition [51, 81, 85, 86] preferring to describe turbulence in terms of the properties it universally displays. Commonly agreed upon properties of turbulent systems include: (i) intrinsic spatio-temporal randomness in the form of chaos, that leads to a sensitive dependence on initial conditions[1]; (ii) a wide range of relevant length and time scales present in the system; (iii) a loss of deterministic predictability due to strong interaction of a large number of degrees of freedom, which generally means that the behaviour of turbulent systems must be predicted statistically; (iv) a complex interplay between individual excitations behaving disorderly, but the statistical distributions of these excitations showing order; (v) turbulence is agreed to be strongly non-linear, out of equilibrium, none memoryless (current state depends on past state not only current state), and causal (current state does not depend on future state); (vi) the curl of the velocity field of the transport medium is generally understood to be non-zero but this not always the case as with weak-wave turbulence [87].

1.3 Two-Dimensional Turbulence

Dimensionality plays an important role in most physical phenomena including turbulence. In fact prior to the 1960s, two-dimensional turbulence was considered by many to be impossible to realize in real physical systems. However, landmark works by Kraichnan [88] and Batchelor [89], opened up the field and it began to be taken more seriously. Even then the primary motivation for studying two-dimensional turbulence was to realize a reduced problem to its three-dimensional counter part [78, 90] whereby studying the less complex system would help with the understanding of the more practical three-dimensional counterpart. However given that two-dimensional turbulence has some analytical solutions and is more tractable. It is a lot less computationally expensive to simulate numerically. We now know that two-dimensional

[1] It is worth noting that chaos does not imply turbulence.

turbulence is not only physically realizable but plays a extensive role in practical applications. From a strict point of view there are no purely two-dimensional classical fluids since they all have a certain spatial extent in the third dimension. Like many other physical systems, decreasing the spatial extent in one direction leads to a system where the energy scale for excitations in the transverse direction become prohibitively large compared to excitations in the plane, thereby 'freezing' the system in the transverse direction leading to quasi-two-dimensional systems [86, 91]. It might seem like these systems are somewhat artificial and would only ever occur in laboratory conditions, but many geophysical and astrophysical phenomenon are quasi-two-dimensional. Examples include tightly confined plasmas [62], soap films [92–96], graphene electrical conduction [97], planetary atmosphere [66, 67] and electrically conducting fluids [78, 98].

Although two-dimensional turbulence started as a curiosity, it soon became a field of its own. Progress in two-dimensional fluids broadened the understanding of three-dimensional turbulence, but also revealed notable behaviour that are unique to two-dimensional systems, such as the importance of large scale coherent structures. A famous example of this phenomena, which has persisted for centuries, is Jupiter's Great Red Spot [69, 99, 100]. One would naturally expect similar behaviours in two-dimensional systems as in three-dimensional systems where the large scale excitations decay into incoherent excitations of smaller length scale while conserving the total energy of the system and increasing the entropy. But for two-dimensional system the system instead preferentially transports energy from small scale incoherent flows to large scale coherent flow structures that are long lived due to suppression of viscosity. For certain conditions these states tend to have a decreasing entropy with increasing energy contained in the structured flow which can lead to an effective negative absolute Boltzmann temperature for the turbulent system[2] [101]. These structured flow states are often termed "Onsager vortices" after Nobel Laureate in chemistry Lars Onsager who first predicted the negative absolute temperature phenomenon in two-dimensional turbulence [101]. This phenomenon is now thought to lie at the heart of large-scale weather systems such as cyclones [102] and the Atlantic Ocean Gulf Stream, and the geophysical turbulence of the Earth's mantle and oceans [103]. Negative temperatures are thought to appear in any system where the high energy states have higher occupation than low energy states. Nuclear spin systems [104, 105] and motional degree of freedom in an optical lattice [106] are two systems in which this phenomena was observed. Discussed in Chap. 6 is use of quantum turbulence in superfluid system to observe "Onsager vortices" verifying Onsager's seventy-year-old predictions. This work shows great extension of capabilities to control and observe the dynamics of quantum fluids, which have allowed their properties to be explored in new regimes far from what was previously possible [74, 107, 108].

[2]The turbulent system has a negative temperature in the sense that the entropy decreases as energy is added, but the system only contains the quasi-particles of vorticity. If one considers the fluid as a whole, then there is no breaking of the second law of thermodynamics and the entropy of the universe is indeed increasing.

1.4 Quantum Turbulence

In the past three decades quantum turbulence (QT) has become a major field of research in superfluid systems. It was first seriously studied in superfluid helium in the 1990s [109–111] even though the medium had been studied since the 1930s. Later, quantum turbulence was observed in Bose-Einstein condensates [112] and ultracold Fermi gasses [113]. Quantum turbulence is made up of disordered quantum vortex filaments which, due to quantum mechanical effects, can only sustain vorticity (circular flows) of quantized strength given by multiples of $\pm h/m$. The flow surrounds a core where no superfluid is present due to phase coherence considerations, analogous to the vortex that forms in flushed toilets where the fluid flows around the edges but is missing from the center. The core of these quantum vortex filaments has width $\sim$1 Å for Helium and $\sim$1 μm for Bose-Einstein condensates of cold atoms. The quantum vortices span the width of the quantum fluid with highly localized velocity fields [114].

Quantum turbulence is believed to arise in the core of neutron stars [115–117], pulsar glitches [118], and even in the quark-gluon plasma [119, 120] and thought to have played a role in the early evolution of the universe, a few microseconds after the Big Bang [121]. To take the example of a pulsar star, from the radiation they emit, it has been observed that their rotation slows down as expected from energy dissipation from the system. However, remarkable glitches are observed where the rotation instantaneously speeds up again. A leading hypothesis ties these glitches into dynamics of quantised vortices in the neutron star core. As well as finding applications in understanding the physical world, many technologies are starting to use superfluid helium as coolant. The understanding of quantum turbulence dynamics is paramount for these applications since it informs the maximum energy transfer rate. Examples of these applications include magnetic resonance imaging (MIR) [122], nuclear magnetic resonance (NMR), magnetoencephalography (MEG), and experiments in physics, such as infrared astronomy device, low temperature Mössbauer spectroscopy [123], particle accelerators, and fusion experiments [124].

The study of quantum turbulence was first described by Feynman [125] who suggested a mechanism for the dissipation of energy in viscosity free environment through the reconnection of quantum vortex tangles, which form vortex rings and that might decay into smaller rings transporting energy from large scale to small scales analogous to what usually happens in 3D classical turbulence. Early experiments studied the dual fluid nature of quantum turbulence in helium, due to the presence of a non-superfluid (normal) fraction of liquid helium overlapped with superfluid (inviscid) component, and tested Feynman's description of turbulence and showed that it could explain mutual friction of a heat current through superfluid ^{4}He [126–128]. With more than 50 years of study in superfluid helium, quantum turbulence is a well established field [109, 114, 124, 129–131]. One might assume that since turbulence in quantum systems is localized, due to quantum mechanical constraints [132–136], the problem should be theoretically simpler than classical turbulence which is thought to be described by Navier-Stokes equations [53]. This is in fact not the case, due to strong interactions with the normal fluid component, and liquid helium still does not have an agreed upon model that accurately describes turbulence at a microscopic level. Due to the small size of the vortex cores which make turbulence hard to image in superfluid helium, and the complications that come from having a strong interactions, studies have mostly focussed on macroscopic quantities such as the transport of energy between large and small scales [137], the rate of turbulent energy dissipation due to thermal friction [126–128], and turbulence coupling to sound modes for the purpose of correction in superfluid helium inertial sensors [74]. Direct measurement of the dissipation mechanisms and dynamics of quantum turbulence in superfluid helium is highly challenging. Instead phenomenological models are tested which include vortex filament models [138], two-fluid models [139] or the Gross-Pitaevskii equation [140] and its modifications [141].

In the past decade dilute-atomic-gas Bose-Einstein condensates have been demonstrated to exhibit quantum turbulence and offer a more versatile environment for the direct study of turbulence [107, 108, 142–146]. Bose-Einstein condensates are a somewhat recent addition to the experimental tool box with their first realization achieved in 1995 [147, 148]. BECs are frequently made using alkali atoms with ^{87}Rb and ^{23}Na. The atom number in the condensates formed using laser cooling ranges from $\sim 10^3$–$\sim 10^7$ [149, 150] and up to 10^9 in the case of hydrogen which uses an initial cryogenic trap [151]. BECs are typically trapped, shaped and probed using magnetic and optical fields and can reach temperatures down to the nanokelvin range. As previously mentioned, for atomic BECs the typical diameter of the vortex core when working with quantum turbulence is on the order of $\sim 1\ \mu$m which is an achievable optical resolution. This allows for the use of light to generate spatial distributions of quantum turbulence [107, 108, 152–154] and also for the in situ imaging of vortices [155] allowing for direct study of quantum turbulence, such as the direct observation of vortex-anitvortex annihilation [156]. When studying quantum turbulence with BECs, there is the extra advantage of the superfluid atoms and the thermal atoms interacting weakly which simplifies the models needed to simulate the behaviour of the system. The non-linear Schrödinger equation, or Gross-Pitaevski equation [157], can be used to obtain accurate numerical simulations of the expected microscopic behaviour of BECs. Lastly, the high level of control over optical and magnetic trapping potentials means that quantum turbulence can be studied in three-dimensions [143, 145] and two-dimensions [107, 156, 158–161] and the interactions between the constituents particles can be tuned to control the energy in the system [158, 162, 163]. Additionally complexity such as disordered potentials [164], long range (dipolar) interactions [165], or multiple internal states [166–168] can be implemented in the system if needed. The combination of accurate numerics, versatile and controllable experimental environment have made BECs an attractive proposition for the study of far-from equilibrium phenomena such as quantum turbulence.

1.5 Two-Dimensional Quantum Turbulence

Due to the ease with which arbitrary 2D systems can be created and quantum turbulence can be observed and manipulated in BECs it became the system of choice for our studies. While very recent progress has been made in the production and investigation of two-dimensional turbulence in superfluid helium [74], it still cannot be achieved and studied easily in these systems due to the technical difficulties associated with achieving appropriate confinement and complex experimental techniques needed to perform meaningful measurement of the vortex distribution. As discussed previously this is not the case for dilute Bose-Einstein condensed gasses where two-dimensional quantum turbulence has already been achieved and directly observed [107, 108, 156, 159–161]. Two-dimensional quantum turbulence is the least complex known form of quantum turbulence, where all the extra vortex excitations have been frozen out and the vortices are confined to move in a plane, and only the position of the straight-line quantum vortex cores in the plane is left, leading to a low number of degrees of freedom.

Theoretical links between two-dimensional classical and 2DQT have already been studied using the Gross-Pitaevski equation and have found non-trivial similarities such as the existence of the von-Karman vortex street [169], an equivalence to the Reynolds number in superfluid systems [170], and superfluid boundary layers [171], as well as evaporative heating and existence of a negative temperature regime [172] similar to regime predicted by Onsager. Similar links have been found between three-dimensional quantum turbulence and classical turbulence which shows that a better understanding of quantum turbulence should lead to a deeper understanding of turbulence as a whole while providing a more complete picture of the turbulence phenomena by uncovering its universal features.

Over the past two decades, the tools to create quantum turbulence have greatly evolved and have been closely liked to the evolution of novel trapping methods. Over the past five years novel spatial light modulator trapping methods, discussed in details in Chap. 4, have made it possible to experimentally study the similarities between quantum turbulence and classical turbulence by using conventional means, such as propellers, oval barriers, and cylinder barrier grids, to generate turbulence in two-dimensional quantum systems. See Ref. [144] for a review of BEC vortex experiments up to 2010. Very recently von-Karman vortex street was observed in atomic BECs [173]. But for most 2DQT studies progress is partially hindered due to difficulties in reaching steady state of turbulence. These difficulties are usually caused by the decay of turbulence and unwanted dynamics driven by external factor such as density gradients in the fluid.

1.6 Scope of This Thesis

The purpose of this thesis is to present a system in which superfluid transport and two-dimensional quantum turbulence can be studied. The system is used to develop models that will be the foundational principles for designing superfluid circuits in the future. It is also used to establish dilute-gas Bose-Einstein condensates as the prototypical system for the experimental study of point vortices and their non-equilibrium dynamics.

More specifically we present our experimental, numerical and theoretical studies of transport and turbulence in oblate superfluid Bose-Einstein condensates, trapped in our newly developed versatile atomic trap generator using the projection of a digital micromirror device (DMD). In addition to describing our trapping techniques and their applicability to multiple experiments, we focus on transport in a tunable superfluid circuit, and explore fundamental aspects of two-dimensional quantum turbulence, observing Onsager's well-known vortex clusters for the first time. The ever-increasing control of such superfluid systems through our work with DMDs has enabled these studies, and these superfluid systems are fundamentally interesting as they present intersections between quantum mechanics and classical fluid dynamics.

The experimental study of superfluid turbulence and transport in BECs is still in the early stages of development. As the experimental techniques have needed time to advance, most of the studies in the quantum turbulence in BECs have been numerical. With improving experimental techniques, driven in part by the increased interest in creating state of the art sensors using BECs, this status appears to be changing. Chapter 4 is fully dedicated to the intricacies and fundamental techniques for using DMDs for the purpose of creating nearly arbitrary two-dimensional optical potentials for the purposes of studying high-resolution dynamics in BECs. This chapter will be of interest to an experimentalist looking to recreate the methods used in this thesis to further their own research.

Chapter 5 presents our work on turbulence and transport in superfluid circuits. Reflecting the challenges of standard electronics, superfluid circuits involve many-body interactions in open systems out of equilibrium. Much of what goes on at the microscopic scale can be averaged out using statistical mechanics which allows theoretical descriptions that treat the system as a fully quantum circuit evolving under unitary evolution [21, 174–179]. Complicated elements like transistors require an open system approach for their gain [180–185] and heat exchanges can be treated as a form of information of the system [186–190]. Superfluids used to process information in a circuit are usually isolated and out of equilibrium from their thermal environment, whereas electrons are strongly coupled to their environment. This decoupling means that the system is not fully a quantum system due to the need to model the perturbation caused by the thermal component on the superfluid, nor can the superfluid be treated as being in thermal equilibrium with its environment. Despite the microscopic complexity of these systems, it is possible to develop lumped-element abstraction models as is done to the electrical circuits to model the transport of superfluid in simple circuits [191]. Here we develop a new approach to superfluid circuitry based on classical acoustic circuits, demonstrating its conceptual and quantitative superiority over previous lumped-element models. These models were developed with the aim of establishing foundational principles of superfluid circuitry that will impact the design of future transport experiments and new generation quantum devices, such as atomtronics circuits and superfluid sensors.

In the 70 years since Onsager's landmark work, his prediction that large-scale structure formation arises from two-dimensional vortices has never before been experimentally tested or observed. In Chap. 6, we put Onsager's theory to the test using two-dimensional Bose-Einstein condensates of atoms to, for the first time, experimentally determine if negative temperature states are physically possible and if so investigate whether they are stable as predicted. We observe long lived large-scale flow in non-driven systems which are characterized by an negative-absolute-Boltzmann temperature. These results demonstrate that dilute-gas BECs are powerful platforms to explore quantum turbulence going forward in the future.

1.7 Outline

The thesis is structured as follows. In Chap. 2 we introduce the Bose-Einstein condensation phenomenon and much of the background theory for quantum fluids and quantum turbulence through the Gross-Pitaevskii equation. Chapter 2 also introduces the theory behind optical atom manipulation and the three main data acquisition techniques used to acquire information about the BEC which are: absorption imaging, Faraday imaging, and momentum spectroscopy using Bragg scattering. Chapter 3 gives an overview of the apparatus used for the experiments presented in other chapters. Chapter 4 presents original work on the utilization of directly imaged DMDs to broaden the control over the potentials available for BEC experiments, which enables the experiments presented in other chapters. Chapter 5 presents our work on studying superfluid transport in the framework of trying to model them using elements of classical transport inspired by [191]. Chapter 6 presents our realization of high-energy Onsager vortex clusters in a planar superfluid and our subsequent study of their stability and dynamics. Chapter 7 presents the future experimental outlook and concluding remarks.

References

1. Bird RB (2002) Transport phenomena. Appl Mech Rev 55:R1–R4
2. Dematteis G, Grafke T, Onorato M, Vanden-Eijnden E (2019) Experimental evidence of hydrodynamic instantons: the universal route to rogue waves. Phys Rev X 9:041057
3. Fu B, Chen M (2020) Linear non-equilibrium thermal and deformation transport model for hematite pellet in dielectric and magnetic heating. Appl Therm Eng 115197
4. Kumar S (2019) Special issue on advances in quantum transport. J Phys Condens Matter 31:200301

5. Batchelor CK, Batchelor G (2000) An introduction to fluid dynamics. Cambridge University Press, Cambridge
6. Toro EF (2013) Riemann solvers and numerical methods for fluid dynamics: a practical introduction. Springer Science & Business Media
7. Fick A (1855) Ueber diffusion. Annalen der Physik 170:59–86
8. Vaishnav J, Ruseckas J, Clark CW, Juzeliunas G (2008) Spin field effect transistors with ultracold atoms. Phys Rev Lett 101:265302
9. Caliga SC, Straatsma CJ, Zozulya AA, Anderson DZ (2016) Principles of an atomtronic transistor. New J Phys 18:015012
10. Caliga SC, Straatsma CJ, Anderson DZ (2017) Experimental demonstration of an atomtronic battery. New J Phys 19:013036
11. Brantut J-P, Meineke J, Stadler D, Krinner S, Esslinger T (2012) Conduction of ultracold fermions through a mesoscopic channel. Science 337:1069–1071
12. Lee JG, McIlvain BJ, Lobb CJ, Hill WTI (2013) Analogs of basic electronic circuit elements in a free-space atom chip. Sci Rep 3:1034
13. Krinner S, Esslinger T, Brantut J-P (2017) Two-terminal transport measurements with cold atoms. J Phys Condens Matter 29:343003
14. Albiez M et al (2005) Direct observation of tunneling and nonlinear self-trapping in a single bosonic Josephson junction. Phys Rev Lett 95:010402
15. Valtolina G et al (2015) Josephson effect in fermionic superfluids across the BEC-BCS crossover. Science 350:1505
16. Spagnolli G et al (2017) Crossing over from attractive to repulsive interactions in a tunneling bosonic Josephson junction. Phys Rev Lett 118:230403
17. Burchianti A et al (2017) Connecting dissipation and phase slips in a Josephson junction between fermionic superfluids. ArXiv e-prints
18. Krinner S, Stadler D, Husmann D, Brantut J-P, Esslinger T (2015) Observation of quantized conductance in neutral matter. Nature 517:64–67
19. Li A et al (2016) Superfluid transport dynamics in a capacitive atomtronic circuit. Phys Rev A 94:023626
20. Piazza F, Collins LA, Smerzi A (2013) Critical velocity for a toroidal Bose-Einstein condensate flowing through a barrier. J Phys B: At Mol Opt Phys 46:095302
21. Eckel S et al (2014) Hysteresis in a quantized superfluid 'atomtronic' circuit. Nature 506:200
22. Fang J, Qin J (2012) Advances in atomic gyroscopes: a view from inertial navigation applications. Sensors 12:6331–6346
23. Barrett B, Bertoldi A, Bouyer P (2016) Inertial quantum sensors using light and matter. Physica Scripta 91:053006
24. Peters A, Chung KY, Chu S (1999) Measurement of gravitational acceleration by dropping atoms. Nature 400:849–852
25. Kuhn C et al (2014) A Bose-condensed, simultaneous dual-species Mach-Zehnder atom interferometer. New J Phys 16:073035
26. Asenbaum P et al (2017) Phase shift in an atom interferometer due to spacetime curvature across its wave function. Phys Rev Lett 118:183602
27. Tino G, Vetrano F (2007) Is it possible to detect gravitational waves with atom interferometers? Class Quantum Gravity 24:2167
28. Dimopoulos S, Graham PW, Hogan JM, Kasevich MA, Rajendran S (2008) Atomic gravitational wave interferometric sensor. Phys Rev D 78:122002
29. Norcia MA, Cline JR, Thompson JK (2017) Role of atoms in atomic gravitational-wave detectors. Phys Rev A 96:042118
30. Graham PW, Hogan JM, Kasevich MA, Rajendran S (2016) Resonant mode for gravitational wave detectors based on atom interferometry. Phys Rev D 94:104022
31. Xu V et al (2019) Probing gravity by holding atoms for 20 seconds. Science 366:745–749
32. Abbott BP et al (2016) Observation of gravitational waves from a binary black hole merger. Phys Rev Lett 116:061102

33. Abbott BP et al (2017) GW170817: observation of gravitational waves from a binary neutron star inspiral. Phys Rev Lett 119:161101
34. Kandes M, Carretero-Gonzalez R, Bromley M (2013) Phase-shift plateaus in the sagnac effect for matter waves. arXiv:13061308
35. Ragole S, Taylor JM (2016) Interacting atomic interferometry for rotation sensing approaching the Heisenberg Limit. Phys Rev Lett 117:203002
36. Stevenson R, Hush MR, Bishop T, Lesanovsky I, Fernholz T (2015) Sagnac interferometry with a single atomic clock. Phys Rev Lett 115:163001
37. McDonald GD et al (2014) Bright solitonic matter-wave interferometer. Phys Rev Lett 113:013002
38. Helm J, Cornish S, Gardiner S (2015) Sagnac interferometry using bright matter-wave solitons. Phys Rev Lett 114:134101
39. Halkyard P, Jones M, Gardiner S (2010) Rotational response of two-component Bose-Einstein condensates in ring traps. Phys Rev A 81:061602
40. Nolan SP, Sabbatini J, Bromley MW, Davis MJ, Haine SA (2016) Quantum enhanced measurement of rotations with a spin-1 Bose-Einstein condensate in a ring trap. Phys Rev A 93:023616
41. Szigeti SS, Lewis-Swan RJ, Haine SA (2017) Pumped-up SU (1, 1) interferometry. Phys Rev Lett 118:150401
42. Luo C, Huang J, Zhang X, Lee C (2017) Heisenberg-limited Sagnac interferometer with multiparticle states. Phys Rev A 95:023608
43. Helm JL, Billam TP, Rakonjac A, Cornish SL, Gardiner SA (2018) Spin-orbit-coupled interferometry with ring-trapped Bose-Einstein condensates. Phys Rev Lett 120:063201
44. Wang Y-J et al (2005) Atom Michelson interferometer on a chip using a Bose-Einstein condensate. Phys Rev Lett 94:090405
45. Garcia O, Deissler B, Hughes K, Reeves J, Sackett C (2006) Bose-Einstein-condensate interferometer with macroscopic arm separation. Phys Rev A 74:031601
46. Burke J, Sackett C (2009) Scalable Bose-Einstein-condensate Sagnac interferometer in a linear trap. Phys Rev A 80:061603
47. Bell TA (2020) Engineering time-averaged optical potentials for Bose-Einstein condensates
48. Meuel T et al (2013) Intensity of vortices: from soap bubbles to hurricanes. Sci Rep 3:3455
49. Batchelor GK (1953) The theory of homogeneous turbulence. Cambridge University Press, Cambridge
50. Tennekes H, Lumley JL (1972) A first course in turbulence. MIT press, Cambridge
51. Frost W (1977) Handbook of turbulence: volume 1 fundamentals and applications. Plenum Press, New York
52. Lesieur M (1987) Turbulence in fluids: stochastic and numerical modelling. Nijhoff Boston, MA
53. Frisch U (1995) Turbulence: the legacy of A. N. Kolmogorov. Cambridge University Press, Cambridge
54. Pope SB (2001) Turbulent flows. Cambridge University Press, Cambridge
55. Ikeda K, Daido H, Akimoto O (1980) Optical turbulence: chaotic behavior of transmitted light from a ring cavity. Phys Rev Lett 45:709
56. McLaughlin D, Moloney JV, Newell AC (1985) New class of instabilities in passive optical cavities. Phys Rev Lett 54:681
57. Akhmanov S, Vorontsov M, Ivanov V (1988) Interactions non linéaires transversales à grande échelle dans les faisceaux de laser; nouveaux types d'ondes non linéaires, apparition de la turbulence optique. Pis ma v žurnal èksperimental noj i teoretičeskoj fiziki 47:611-614
58. Aceves A (1993) in Nonlinear waves and weak turbulence 199–210. Springer, Berlin
59. Grappin R, Frisch U, Pouquet A, Leorat J (1982) Alfvenic fluctuations as asymptotic states of MHD turbulence. Astron Astrophys 105:6–14
60. Garbet X (2001) Turbulence in fusion plasmas: key issues and impact on transport modelling. Plasma Phys Control Fusion 43:A251

61. Monchaux R et al (2007) Generation of a magnetic field by dynamo action in a turbulent flow of liquid sodium. Phys Rev Lett 98:044502
62. Conway G (2008) Turbulence measurements in fusion plasmas. Plasma Phys Control Fusion 50:124026
63. Kunz MW, Abel IG, Klein KG, Schekochihin AA (2018) Astrophysical gyrokinetics: turbulence in pressure-anisotropic plasmas at ion scales and beyond. J Plasma Phys 84:715840201
64. Warhaft Z (2002) Turbulence in nature and in the laboratory. Proc Natl Acad Sci 99:2481–2486
65. Wlazlowski G, Sekizawa K, Magierski P, Bulgac A, Forbes MM (2016) Vortex pinning and dynamics in the neutron star crust. Phys Rev Lett 117:232701
66. Lindborg E (1999) Can the atmospheric kinetic energy spectrum be explained by two-dimensional turbulence? J Fluid Mech 388:259–288
67. Boffetta G, Ecke RE (2011) Two-dimensional turbulence. Ann Rev Fluid Mech 44:427–451
68. Williams PD, Joshi MM (2013) Intensification of winter transatlantic aviation turbulence in response to climate change. Nat Clim Chang 3:644
69. Marcus PS (1988) Numerical simulation of Jupiter's great red spot. Nature 331:693–696
70. Yamazaki H, Mitchell JG, Seuront L, Wolk F, Li H (2006) Phytoplankton microstructure in fully developed oceanic turbulence. Geophys Res Lett 33
71. Durham WM et al (2013) Turbulence drives microscale patches of motile phytoplankton. Nat Commun 4:1–7
72. De Lillo F et al (2014) Turbulent fluid acceleration generates clusters of gyrotactic microorganisms. Phys Rev Lett 112:044502
73. Hu DL, Chan B, Bush JWM (2003) The hydrodynamics of water strider locomotion. Nature 424:663–666
74. Sachkou YP et al (2019) Coherent vortex dynamics in a strongly interacting superfluid on a silicon chip. Science 366:1480–1485
75. Sachkou Y (2019) Probing two-dimensional quantum fluids with cavity optomechanics
76. Ruelle D, Takens F (1971) On the nature of turbulence. Les rencontres physiciens-mathématiciens de Strasbourg-RCP25 12:1–44
77. Falkovich G, Sreenivasan KR (2006) Lessons from hydrodynamic turbulence Technical Report. Abdus Salam International Centre for Theoretical Physics
78. Tabeling P (2002) Two-dimensional turbulence: a physicist approach. Phys Rep 362:1–62
79. Hof B, de Lozar A, Avila M, Tu X, Schneider TM (2010) Eliminating turbulence in spatially intermittent flows. Science 327:1491
80. Van den Akker HE (2006) The details of turbulent mixing process and their simulation. Adv Chem Eng 31:151–229
81. Han Z, Reitz RD (1995) Turbulence modeling of internal combustion engines using RNG $nkappa\text{-}nvarepsilon$ models. Combust Sci Technol 106:267–295
82. Glassman I, Yetter RA Glumac NG (2014) Combustion. Academic press
83. Crabtree GW, Nelson DR (1997) Vortex physics in high-temperature superconductors. Phys Today 50:38–45
84. Field B, Simula T (2018) Introduction to topological quantum computation with non-Abelian anyons. Quantum Sci Technol 3:045004
85. Tsinober A (2001) An informal introduction to turbulence. Kluwer Academic Publishers, Dordrecht
86. Davidson PA (2015) Turbulence: an introduction for scientists and engineers. Oxford University Press, Oxford
87. Nazarenko S (2011) Wave turbulence. Springer Science & Business Media
88. Kraichnan RH (1967) Inertial ranges in two-dimensional turbulence. Phys Fluids 10:1417–1423
89. Batchelor GK (1969) Computation of the energy spectrum in homogeneous two-dimensional turbulence. Phys Fluids 12:II-233
90. Shats M, Xia H, Punzmann H (2005) Spectral condensation of turbulence in plasmas and fluids and its role in low-to-high phase transitions in toroidal plasma. Phys Rev E 71:046409
91. Kraichnan RH, Montgomery D (1980) Two-dimensional turbulence. Rep Prog Phys 43:547

92. Gharib M, Derango P (1989) A liquid film (soap film) tunnel to study two-dimensional laminar and turbulent shear flows. Phys D: Nonlinear Phenom 37:406–416
93. Martin B, Wu X, Goldburg W, Rutgers M (1998) Spectra of decaying turbulence in a soap film. Phys Rev Lett 80:3964
94. Rutgers MA (1998) Forced 2D turbulence: experimental evidence of simultaneous inverse energy and forward enstrophy cascades. Phys Rev Lett 81:2244–2247
95. Vorobieff P, Ecke RE (1999) Cylinder wakes in flowing soap films. Phys Rev E 60:2953
96. Wen C-Y, Lin C-Y (2001) Two-dimensional vortex shedding of a circular cylinder. Phys Fluids 13:557–560
97. Gupta KS, Sen S (2010) Turbulent flow in graphene. EPL (Europhys Lett) 90:34003
98. Kellay H, Goldburg WI (2002) Two-dimensional turbulence: a review of some recent experiments. Rep Prog Phys 65:845
99. Sommeria J, Meyers SD, Swinney HL (1988) Laboratory simulation of Jupiter's Great Red Spot. Nature 331:689–693
100. Bouchet F, Sommeria J (2002) Emergence of intense jets and Jupiter's Great Red Spot as maximumentropy structures. J Fluid Mech 464:165–207
101. Onsager L (1949) Statistical hydrodynamics. Il Nuovo Cimento 1943–1954(6):279–287
102. Cannon J, Shivamoggi B (2006) Mathematical and physical theory of turbulence. CRC Press
103. Van Atta C, Chen W (1970) Structure functions of turbulence in the atmospheric boundary layer over the ocean. J Fluid Mech 44:145–159
104. Purcell EM, Pound RV (1951) A nuclear spin system at negative temperature. Phys Rev 81:279
105. Ramsey NF (1956) Thermodynamics and statistical mechanics at negative absolute temperatures. Phys Rev 103:20
106. Braun S et al (2013) Negative absolute temperature for motional degrees of freedom. Science 339:52–55
107. Gauthier G et al (2019) Giant vortex clusters in a two-dimensional quantum fluid. Science 364:1264–1267
108. Johnstone SP et al (2019) Evolution of large-scale flow from turbulence in a two-dimensional superfluid. Science 364:1267–1271
109. Donnelly RJ (1993) Quantized vortices and turbulence in helium II. Ann Rev Fluid Mech 25:325–371
110. Maurer J, Tabeling P (1998) Local investigation of superfluid turbulence. EPL (Europhys Lett) 43:29
111. Stalp SR, Skrbek L, Donnelly RJ (1999) Decay of grid turbulence in a finite channel. Phys Rev Lett 82:4831
112. Dalfovo F, Giorgini S, Pitaevskii LP, Stringari S (1999) Theory of Bose-Einstein condensation in trapped gases. Rev Mod Phys 71:463
113. Bulgac A, Forbes MM, Wlaznlowski G (2016) Towards quantum turbulence in cold atomic fermionic superfluids. J Phys B: At Mol Opt Phys 50:014001
114. Paoletti MS, Lathrop DP (2011) Quantum turbulence. Ann Rev Condens Matter Phys 2:213–234
115. Peralta C, Melatos A, Giacobello M, Ooi A (2005) Global three-dimensional flow of a neutron superfluid in a spherical shell in a neutron star. Astrophys J 635:1224
116. Peralta C, Melatos A, Giacobello M, Ooi A (2006) Transitions between turbulent and laminar superfluid vorticity states in the outer core of a neutron star. Astrophys J 651:1079
117. Andersson N, Sidery T, Comer GL (2007) Superfluid neutron star turbulence. Mon Not R Astronom Soc 381:747–756
118. Melatos A, Peralta C (2007) Superfluid turbulence and pulsar glitch statistics. Astrophys J 662:L99–L102
119. Davidson M (2010) The quark-gluon plasma and the stochastic interpretation of quantum mechanics. Proc Int Conf Front Quantum Mesoscopic Thermodyn FQMT '08 42:317–322
120. Rogachevskii I et al (2017) Laminar and turbulent dynamos in chiral magnetohydrodynamics. I. Theory. Astrophys J 846:153

121. Yagi K, Hatsuda T, Miake Y (2005) Quark-gluon plasma: From big bang to little bang. Cambridge University Press, Cambridge
122. Brown RW, Cheng Y-CN, Haacke EM, Thompson MR, Venkatesan R (2014) Magnetic resonance imaging: physical principles and sequence design. Wiley
123. Greenwood NN (2012) Mössbauer spectroscopy. Springer Science & Business Media
124. Barenghi CF, Donnelly RJ, Vinen W (2001) Quantized vortex dynamics and superfluid turbulence. Springer Science & Business Media
125. Feynman RP (1955) In: Gorter CJ (ed) Elsevier, pp 17–53
126. Vinen WF (1957) Mutual friction in a heat current in liquid helium II I. Experiments on steady heat currents. Proc R Soc Lond Ser A Math Phys Sci 240:114–127
127. Vinen WF (1957) Mutual friction in a heat current in liquid helium II. II. Experiments on transient effects. Proc R Soc Lond Ser A Math Phys Sci 240:128–143
128. Vinen WF (1957) Mutual friction in a heat current in liquid helium II III. Theory of the mutual friction. Proc R Soc Lond Ser A Math Phys Sci 242:493–515
129. Vinen W (2006) An introduction to quantum turbulence. J Low Temp Phys 145:7–24
130. Barenghi CF, L'vov VS, Roche P-E (2014) Experimental, numerical, and analytical velocity spectra in turbulent quantum fluid. Proc Natl Acad Sci 111:4683–4690
131. Vinen WF, Donnelly RJ (2007) Quantum turbulence-Because superfluid eddies can only be formed from quantized vortex lines, one might expect quantum turbulence to be very different from its classical counterpart. But that's not necessarily so. Phys Today 60:43
132. Nore C, Abid M, Brachet ME (1997) Kolmogorov turbulence in low-temperature superflows. Phys Rev Lett 78:3896–3899
133. Nore C, Abid M, Brachet M (1997) Decaying Kolmogorov turbulence in a model of superflow. Phys Fluids 9:2644–2669
134. Tsubota M (2009) Quantum turbulence: from superfluid helium to atomic Bose-Einstein condensates. Contemp Phys 50:463–475
135. Barenghi CF (1999) Classical aspects of quantum turbulence. J Phys Condens Matter 11:7751
136. Niemela J, Sreenivasan K, Donnelly R (2005) Grid generated turbulence in helium II. J Low Temp Phys 138:537–542
137. Kivotides D (2014) Energy spectra of finite temperature superfluid helium-4 turbulence. Phys Fluids 26:105105
138. Vinen W, Niemela J (2002) Quantum turbulence. J Low Temp Phys 128:167–231
139. Vassilicos J et al (2000) Turbulence structure and vortex dynamics. Cambridge University Press, Cambridge
140. Roberts PH, Berloff NG (2001) Quantized vortex dynamics and superfluid turbulence. Springer, Berlin, pp 235–257
141. Berloff NG, Roberts PH (2001) Quantized vortex dynamics and superfluid turbulence. Springer, Berlin, pp 268–275
142. Donnelly RJ (1991) Quantized vortices in Helium II. Cambridge University Press, Cambridge
143. Henn E, Seman J, Roati G, Magalhaes KMF, Bagnato VS (2009) Emergence of turbulence in an oscillating Bose-Einstein condensate. Phys Rev Lett 103:045301
144. Anderson BP (2010) Resource article: experiments with vortices in superfluid atomic gases. J Low Temp Phys 161:574–602
145. Henn E, Seman J, Roati G, Magalhaes K, Bagnato V (2010) Generation of vortices and observation of quantum turbulence in an oscillating Bose-Einstein condensate. J Low Temp Phys 158:435
146. Seman J et al (2011) Route to turbulence in a trapped Bose-Einstein condensate. Laser Phys Lett 8:691
147. Davis KB et al (1995) Bose-Einstein condensation in a gas of sodium atoms. Phys Rev Lett 75:3969–3973
148. Anderson MH, Ensher JR, Matthews MR, Wieman CE, Cornell EA (1995) Observation of Bose-Einstein condensation in a dilute atomic vapor. Science 269:198
149. Hänsel W, Hommelhoff P, Hänsch TW, Reichel J (2001) Bose-Einstein condensation on a microelectronic chip. Nature 413:498–501

150. Van der Stam KMR, van Ooijen ED, Meppelink R, Vogels JM, van der Straten P (2007) Large atom number Bose-Einstein condensate of sodium. Rev Sci Instrum 78:013102
151. Fried DG et al (1998) Bose-Einstein condensation of atomic hydrogen. Phys Rev Lett 81:3811–3814
152. Gaunt AL, Schmidutz TF, Gotlibovych I, Smith RP, Hadzibabic Z (2013) Bose-Einstein condensation of atoms in a uniform potential. Phys Rev Lett 110:200406
153. Samson E, Wilson K, Newman Z, Anderson B (2016) Deterministic creation, pinning, and manipulation of quantized vortices in a Bose-Einstein condensate. Phys Rev A 93:023603
154. Gauthier G et al (2016) Direct imaging of a digital-micromirror device for configurable microscopic optical potentials. Optica 3:1136–1143
155. Wilson KE, Newman ZL, Lowney JD, Anderson BP (2015) In situ imaging of vortices in Bose-Einstein condensates. Phys Rev A 91:023621
156. Kwon WJ, Moon G, Choi J-Y, Seo SW, Shin Y-I (2014) Relaxation of superfluid turbulence in highly oblate Bose-Einstein condensates. Phys Rev A 90:063627
157. Pethick CJ, Smith H (2008) Bose-Einstein condensation in dilute gases, 2nd edn. Cambridge University Press, Cambridge
158. Bloch I, Dalibard J, Zwerger W (2008) Many-body physics with ultracold gases. Rev Mod Phys 80:885
159. Neely TW et al (2013) Characteristics of two-dimensional quantum turbulence in a compressible superfluid. Phys Rev Lett 111:235301
160. Kim JH, Kwon WJ, Shin Y-I (2016) Role of thermal friction in relaxation of turbulent Bose-Einstein condensates. Phys Rev A 94:033612
161. Seo SW, Ko B, Kim JH, Shin Y (2017) Observation of vortex-antivortex pairing in decaying 2D turbulence of a superfluid gas. Sci Rep 7:4587
162. Inouye S et al (1998) Observation of Feshbach resonances in a Bose-Einstein condensate. Nature 392:151–154
163. Chin C, Grimm R, Julienne P, Tiesinga E (2010) Feshbach resonances in ultracold gases. Rev Mod Phys 82:1225–1286
164. Fallani L, Fort C, Inguscio M (2008) Bose-Einstein condensates in disordered potentials. Adv At Mol Opt Phys 56:119–160
165. Lu M, Burdick NQ, Youn SH, Lev BL (2011) Strongly dipolar Bose-Einstein condensate of dysprosium. Phys Rev Lett 107:190401
166. Hall D, Matthews M, Ensher J, Wieman C, Cornell EA (1998) Dynamics of component separation in a binary mixture of Bose-Einstein condensates. Phys Rev Lett 81:1539
167. Stenger J et al (1998) Spin domains in ground-state Bose-Einstein condensates. Nature 396:345–348
168. Stamper-Kurn DM, Ueda M (2013) Spinor Bose gases: symmetries, magnetism, and quantum dynamics. Rev Mod Phys 85:1191
169. Sasaki K, Suzuki N, Saito H (2010) Bénardnchar21von Kármán Vortex Street in a Bose-Einstein Condensate. Phys Rev Lett 104:150404
170. Reeves MT, Billam TP, Anderson BP, Bradley AS (2015) Identifying a superfluid Reynolds number via dynamical similarity. Phys Rev Lett 114:155302
171. Stagg GW, Parker NG, Barenghi CF (2017) Superfluid boundary layer. Phys Rev Lett 118:135301
172. Simula T, Davis MJ, Helmerson K (2014) Emergence of order from turbulence in an isolated planar superfluid. Phys Rev Lett 113:165302
173. Kwon WJ, Kim JH, Seo SW, Shin Y (2016) Observation of von Kármán vortex street in an atomic superfluid gas. Phys Rev Lett 117:245301
174. Gati R, Albiez M, Fälling J, Hemmerling B, Oberthaler M (2006) Realization of a single Josephson junction for Bose-Einstein condensates. Appl Phys B 82:207–210
175. Gati R, Oberthaler MK (2007) A bosonic Josephson junction. J Phys B: At Mol Opt Phys 40:R61
176. Rab M et al (2008) Spatial coherent transport of interacting dilute Bose gases. Phys Rev A 77:061602

177. Moulder S, Beattie S, Smith RP, Tammuz N, Hadzibabic Z (2012) Quantized supercurrent decay in an annular Bose-Einstein condensate. Phys Rev A 86:013629
178. Ryu C, Blackburn PW, Blinova AA, Boshier MG (2013) Experimental realization of Josephson junctions for an atom SQUID. Phys Rev Lett 111:205301
179. Ryu C, Boshier MG (2015) Integrated coherent matter wave circuits. New J Phys 17:092002
180. Seaman BT, Kramer M, Anderson DZ, Holland MJ (2007) Atomtronics: ultracold-atom analogs of electronic devices. Phys Rev A 75:023615
181. Stickney JA, Anderson DZ, Zozulya AA (2007) Transistorlike behavior of a Bose-Einstein condensate in a triple-well potential. Phys Rev A 75:013608
182. Pepino RA, Cooper J, Anderson DZ, Holland MJ (2009) Atomtronic circuits of diodes and transistors. Phys Rev Lett 103:140405
183. Pepino RA, Cooper J, Meiser D, Anderson DZ, Holland MJ (2010) Open quantum systems approach to atomtronics. Phys Rev A 82:013640
184. Gajdacz M, Opatrny T, Das KK (2014) An atomtronics transistor for quantum gates. Phys Lett A 378:1919–1924
185. Chow WW, Straatsma CJE, Anderson DZ (2015) Numerical model for atomtronic circuit analysis. Phys Rev A 92:013621
186. Rothstein J (1951) Information, measurement, and quantum mechanics. Science 114:171
187. Miller DG (1960) Thermodynamics of irreversible processes. The experimental verification of the onsager reciprocal relations. Chem Rev 60:15–37
188. Landauer R (1961) Irreversibility and heat generation in the computing process. IBM J Res Dev 5:183–191
189. Jendrzejewski F et al (2014) Resistive flow in a weakly interacting Bose-Einstein condensate. Phys Rev Lett 113:045305
190. Faist P, Dupuis F, Oppenheim J, Renner R (2015) The minimal work cost of information processing. Nat Commun 6:7669
191. Eckel S et al (2016) Contact resistance and phase slips in mesoscopic superfluid-atom transport. Phys Rev A 93:063619

Chapter 2
Theoretical Background

This chapter provides the relevant theoretical background necessary for understanding the experiments presented in this thesis. First, details on Bose-Einstein condensation, quantum fluids, and turbulence are presented. The basic properties of the Gross-Pitaevskii equation are introduced as they provide a theoretical model for the zero-temperature dynamics of the BEC; this discussion includes its hydrodynamic form and the resulting quantum vortices. The point vortex model is introduced, as it provides a good approximation of vortex dynamics in quasi-uniform two-dimensional condensates. This chapter concludes by describing the theoretical foundations of the data acquisition techniques used in the experiments, along with a description of the calibration techniques used. Primarily, both Faraday and absorption imaging were used to measure the atomic density distribution and are thus described. Finally, an overview of Bragg-scattering, which was used to perform the momentum spectroscopy of the BEC and for vortex sign detection, is presented.

2.1 Bose-Einstein Condensates

Elementary particles are categorized according to their internal angular momentum with fermions having half-integer spin, and bosons possessing integer spin. The wave functions for these particles are distinguished by their respective symmetries under particle position exchange:

$$\Psi_{\text{boson}}(\vec{r}_1, \vec{r}_2) = \Psi_{\text{boson}}(\vec{r}_2, \vec{r}_1); \qquad \Psi_{\text{fermion}}(\vec{r}_1, \vec{r}_2) = -\Psi_{\text{fermion}}(\vec{r}_2, \vec{r}_1), \tag{2.1}$$

where the overall change of the phase of the wave function for the bosons and fermions evolves by $2n\pi$ and $(2n + 1)\pi$ respectively with $n \in \mathbb{Z}$, and $\vec{r}_1, \vec{r}_2$ represent the position of the particles. These symmetries of the wave functions lead to drastic differences in the behaviour. For bosons, they can occupy the same quantum state, leading to a tendency of the particles to occupy the same state. Meanwhile, fermions

© The Editor(s) (if applicable) and The Author(s), under exclusive license to Springer Nature Switzerland AG 2020

G. Guillaume, *Transport and Turbulence in Quasi-Uniform and Versatile Bose-Einstein Condensates*, Springer Theses,
https://doi.org/10.1007/978-3-030-54967-1_2

cannot occupy the same state, due to the Pauli exclusion principle. These properties lead to very different quantum state occupation statistics for these particles where the bosons follow Bose-Einstein statistics and fermions follow Fermi-Dirac statistics [1].

The development of the boson statistical model arrived in 1924, when Satyendra Nath Bose used the statistics of a collection of quantized particles of light statistics to derive the Plank distribution [2]. In 1924–1925, Einstein used de Broglie's idea about the duality of matter waves to extent Bose's work to massive particles [3, 4] from which the Bose-Einstein probability distribution was born,

$$f^0(E_n) = \frac{1}{e^{E_n - E_{\text{min}}} - 1}. \tag{2.2}$$

This expression reflects the probability of finding a particle in the energy state E_n. Einstein found that bosons which obeyed Bose's proposed framework were indistinguishable from one another which leads to having no limit on the number of particles occupying the same state. Combined with particle conservation, he predicted a critical temperature at which a phase transition starts to occur where the atoms start to preferentially occupy the lowest quantum state of the system. This is a consequence of Eq. (2.2) since the energy of the particles cannot go below the lowest energy state at absolute zero all the atoms must be in the same state, referred to as the Bose-Einstein Condensate (BEC). For particles in this state, the ground state of the system is known as the macroscopic ground state which differs from single particle ground state due to presence of interactions. In the macroscopic ground state of the system, the particles form a coherent, macroscopically populated matter wave which tends to behave as a single macroscopic quantum object which is coherent and behaves as a superfluid. It took 70 years to experimentally realize Einstein's original idea using cold atoms, with the first condensation achieved in 1995 [1].

2.2 Basics of Quantum Fluids

In this section, the basics of quantum fluids are introduced in terms of how they can be modelled and how can they behave. In particular, the Gross-Pitaevskii equation is described, which in most cases provides a good description the behaviour of these fluids at low temperatures. It has been shown to be a good model of ^{87}Rb BECs in the 2D superfluid regime [1] which we achieved experimentally. Quantum vortices and their dynamics will be introduced. They occur in BECs due to the continuity of the macroscopic wave function and form the basis of quantum turbulence (QT) which has emerged as a major field of research in superfluid systems and will be explored in this work. The hydrodynamic modelling of BECs will also be presented which will be used to investigate the transport dynamics of superfluid systems.

2.2.1 Gross-Pitaevskii Equation

As described above, for any ensemble made up of bosons following Bose statistics, there exists a certain critical temperature (T_c) below which the ensemble achieves Bose-Einstein condensation and a large fraction of the particles occupy the macroscopic ground state of the system. In principle, the many-body wave function of a Bose-Einstein condensate is exactly solvable using a many-body Schrödinger equation, but the computational resources required are not practical even for modest atom numbers. Provided that most of the particles are present in the ground state, or that they do not interact with the particles which are in other states of the system, and that the interaction between the particles is not too strong, the Gross-Pitaevskii equation (GPE), first derived in the context of superfluidity as a phenomenological description of superfluid helium [5–7], provides a (semi)classical computationally tractable description of the macroscopic wave function ($\Psi(\vec{r}, t)$) of the system [8]. In this description, the many-body quantum field operator that represents the boson particles in the ensemble is replaced with a (semi)classical field:

$$i\hbar\frac{\partial\Psi(\vec{r}, t)}{\partial t} = \hat{H}\Psi(\vec{r}, t) = \left[\frac{-\hbar}{2m}\vec{\nabla}^2 + V(\vec{r}, t) + g\,|\Psi(\vec{r}, t)|^2\right]\Psi(\vec{r}, t). \tag{2.3}$$

$\hat{H}$ is the Hamiltonian, $V(\vec{r}, t)$ is the net external potential acting on the system, g is the interaction strength parameter of the system, and m is the mass of the constituent particles. If $g > 0$ then the interactions are repulsive since the constituent particles repel each other. Conversely, for $g < 0$ the system is attractive. Lastly, $g = 0$ constitutes an ideal, non-interacting system where the particles are not 'aware' of each other's existence. A non-interacting system obeys the linear Schrödinger equation and the ensemble will behave as a single-particle. The non-linearity in the system thus results from the interactions between the bosons and can be approximated by considering the s-wave scattering, since it is the main collision process at low energy. The coupling constant characterizing the strength of the two-body interaction is given by

$$g = \frac{4\pi\hbar^2 a_s}{m}, \tag{2.4}$$

where a_s is the cross-section of the s-wave interactions for the constituent particles. For systems at higher temperature, higher-order collisions that modify the coupling constant are possible. The derivation of the GPE is not shown here, but it can be found by factorising the many-body wave function to extract the single-particle ground state [9], or by treating the field operator as a linear combination of a mean field and a quantum fluctuation field, and showing that for large enough occupations of the ground state the quantum fluctuations can be ignored [10]. Although the GPE is a semi-classical approximation of the many-body wave function, it captures many aspects of condensate dynamics such as the dynamics of vortices, solitons, sound propagation, and collective oscillations. It does not, however, capture quantum effects such as tunnelling. It also ignores occupations of high energy modes, which occur when not all of the particles are in the condensate state, such as is the case at non-zero temperature in the presence of a thermal cloud.

The experiments presented in this document use condensates of ^{87}Rb which are well described by the GPE since the occupation of the ground state tends to be quite high (>80%) and the interactions are quite small due to the low density of the cloud compared to liquid helium. In the absence of a magnetic field, the scattering cross-section (a_s) of ^{87}Rb is positive (repulsive) and approximately equal to 100 Bohr radii ($a_s = 100a_0$). It can be tuned using Feshbach resonances [11–13] in the presence of a magnetic field, but the experiments in this thesis take place in the presence of small magnetic field ($|\vec{B}| < 80$ G) which has little effect on the scattering cross-section.

For a system consisting of N_{total} with N_{cond} particles in the ground state, the wave function describes the local density throughout the system which gives the following normalization condition for the macroscopic ground state of the system

$$N_{cond} = \int |\Psi(\vec{r}, t)|^2 d^3\vec{r}. \tag{2.5}$$

The energy contained within the system is given by

$$E = \int \left[\frac{\hbar^2}{2m} \left| \vec{\nabla}\Psi(\vec{r}, t) \right|^2 + V(\vec{r}, t) \left| \Psi(\vec{r}, t) \right|^2 + \frac{g}{2} |\Psi(\vec{r}, t)|^4 \right] d^3\vec{r}. \tag{2.6}$$

For an isolated system with a fixed external potential in time, $V(\vec{r}, t) = V(\vec{r}, 0)$, the energy and atom number are conserved which is the case with Eq. (2.3). It is useful to decompose the system into its constituent energies:

$$E_K(t) = \int \frac{\hbar^2}{2m} \left| \vec{\nabla}\Psi(\vec{r}, t) \right|^2 d^3\vec{r}; \tag{2.7a}$$

$$E_P(t) = \int V(\vec{r}, t) \left| \Psi(\vec{r}, t) \right|^2 d^3\vec{r}; \tag{2.7b}$$

$$E_I(t) = \int \frac{g}{2} |\Psi(\vec{r}, t)|^4 d^3\vec{r}; \tag{2.7c}$$

which are the kinetic, potential, and interaction energies respectively.

2.2.2 Time-Independent Gross-Pitaevskii Equation

Stationary states of the GPE can be obtained by separation of variables, assuming that $\hat{H}$ is time independent, as is done for the Schrödinger equation. The stationary solutions have the form $\Psi(\vec{r}, t) = e^{-i\mu t/\hbar}\psi(\vec{r})$ where μ, defined as $\mu \equiv \partial E/\partial N$, is the chemical potential 'eigenvalue' and $\psi(\vec{r})$ is the stationary wave function profile. Substituting into Eq. (2.3) gives the time-independent GPE equation

$$\mu\psi(\vec{r}) = \left[\frac{-\hbar}{2m}\vec{\nabla}^2 + V(\vec{r}) + g\,|\psi(\vec{r})|^2\right]\psi(\vec{r}). \tag{2.8}$$

The chemical potential can be similarly expressed in terms of the system energies as

$$\mu = \frac{1}{N_{\text{cond}}}\left(E_{\text{K}} + E_{\text{P}} + 2E_{\text{I}}\right). \tag{2.9}$$

Cold atom experiments usually start with a condensate in the ground state of the system before modifying the external potential to probe the response of the system. This means that to simulate the initial state of the system, one must first find the stationary state with the lowest energy. This can be achieved by starting with a random initial state that is non-orthogonal to the ground state solution $\Psi_{\text{init}}(\vec{r}, t) = \sum_{i=0}^{\infty} a_i e^{-i\mu_i t/\hbar}\psi(\vec{r})$. Through the use of imaginary time evolution of the wave function [14], $\Psi(\vec{r}, -it) = \sum_{i=0}^{\infty} a_i e^{-\mu_i t/\hbar}\psi(\vec{r})$, in the limit $(t \to \infty)$, the mode with the lowest chemical potential will become dominant. After renormalization, one obtains a very good estimate of the ground state wave function. It is not necessary to decompose the original wave function explicitly, since doing so would mean that the ground state wave function is already known. Instead, one makes the imaginary time substitution in Eq. (2.3) and then evolves an initial ansatz wave function under imaginary time to obtain the ground state. For the initial ansatz ground state, the Thomas-Fermi ground state approximation is usually used, see Sect. 2.2.4.

2.2.3 A Generalized Dimensional Reduction

In most of our experiments, the trapping is highly anisotropic, such that the trapping in the vertical direction is much more tightly confined that in the horizontal directions, with trapping frequencies ranging from $\omega_z = 2\pi \times (140 - 320)$ Hz vertically and $\omega_x \approx \omega_y = 2\pi \times (2.7 - 6.2)$ Hz, horizontally. Note the range results from the choice in the input power of the trapping beam. This imbalance in the frequencies in the trapping leads to a separation in the energy scales between the directions because excitation along different directions will now have different energy. This leads to most and in some cases all of the excitations to occur in the horizontal directions where it is less energy costly, leading to higher entropy. This leads to the possibility of expressing the wave function as products of a 2D density profile $\Psi(x, y, t)$ with a time-independent vertical density profile $\phi(x, y, z)$. The x and y dependence of the vertical ground state is to account for the fact that the vertical trapping frequency in experimental traps is not actually constant, so the vertical trapping potential has the form $V_z(x, y, z)$. The macroscopic wave function can, therefore, be written as $\Psi(\vec{r}, t) = \Psi(x, y, t)\phi(x, y, z)$ with $\int|\phi(x, y, z)|^2 = 1$, $\int|\Psi(x, y, t)|^2 = N$ and the potential as $V(\vec{r}, t) = V(x, y, t) + V_z(x, y, z)$. Note it is possible to include a time dependence on the vertical trapping potential as long as it is fast enough to be time-averaged [15], in which case $V_z(x, y, z)$ would be the time-averaged effective potential. Substituting into the GPE, Eq. (2.3), multiplying by $\phi^*(x, y, z)$, and then integrating the z-dependence results in an effective 2D GPE of the form:

$$i\hbar\frac{\partial\Psi(x, y, t)}{\partial t} = \frac{-\hbar}{2m}\left(\vec{\nabla}^2 + \int dz\phi^*(x, y, z)\vec{\nabla}^2\phi(x, y, z)\right)\Psi(x, y, t) + V_{2D}(x, y, t)\Psi(x, y, t) + g_{2D}(x, y)\left|\Psi(x, y, t)\right|^2\Psi(x, y, t), \tag{2.10}$$

where $V_{2\mathrm{D}}(x, y, t)$ is the effective 2D potential of the 3D system, $g_{2\mathrm{D}}(x, y)$ is the effective interaction strength. These parameters are given by

$$V_{2\mathrm{D}}(x, y, t) = V(x, y, t) + \int dz V_z(x, y, z)|\phi(x, y, z)|^2 = \frac{\int dz V(\vec{r}, t)|\Psi(\vec{r}, t)|^2}{\int dz|\Psi(\vec{r}, t)|^2} \tag{2.11}$$

and

$$g_{2\mathrm{D}}(x, y) = g\int dz|\phi(x, y, z)|^4 = g\frac{\int dz|\Psi(\vec{r}, t)|^4}{\left(\int dz|\Psi(\vec{r}, t)|^2\right)^2} \tag{2.12}$$

where we have used the fact that $\int dz|\phi(x, y, z)|^2 = 1$ and that $\Psi(x, y, t)$ has no z-dependence to make the change from $\phi(x, y, z)$ to $\Psi(\vec{r}, t)$. The reason for this transformation is that the exact functional form of the vertical confinement might be unknown, making it hard to calculate $\phi(x, y, z)$ explicitly. This is not an issue numerically, since one only needs to compute the ground state of the potential $\Psi_0(\vec{r})$ and apply Eqs. (2.11) and (2.12) to the effective potential and interaction strengths. Note that, if possible, it is better to find the ground state of $V_z(x, y, z)$ and not of $V(\vec{r}, 0)$. Even though theoretically there is no difference, numerically the integrals are inaccurate if the local wave function is too small due to $V(x, y, z)$ being high. Finally, there is a $\int dz\phi^*(x, y, z)\vec{\nabla}^2\phi(x, y, z)$ factor in Eq. (2.10) which cannot be trivially calculated. Examining Eq. (2.8), it is equivalent to making the replacement

$$\mu \rightarrow \mu - \frac{\hbar}{2m}\int dz\phi^*(x, y, z)\vec{\nabla}^2\phi(x, y, z). \tag{2.13}$$

In which case Eq. (2.10) becomes simply

$$i\hbar\frac{\partial\Psi(x, y, t)}{\partial t} = \left[\frac{-\hbar}{2m}\vec{\nabla}^2 + V_{2\mathrm{D}}(x, y, t) + g_{2\mathrm{D}}(x, y)\left|\Psi(x, y, t)\right|^2\right]\Psi(x, y, t). \tag{2.14}$$

Unless stated otherwise, all the effective-2D simulations in this thesis are subject to this dimensional reduction. Note that this reduction is not limited to harmonic trapping or system where the atoms are in the vertical ground state. As long as the z trapping potential is independent of time, and the wave function can be decomposed into a 2D fraction, multiplied by a time-independent spatially dependent vertical component this reduction applies. This approach has been extended by Thomas Bell in the UQ lab to simulate the time-of-flight expansion of a BEC [15] where $V(\vec{r}, t) = 0$ and $\Psi(\vec{r}, t) = \Psi(x, y, t)\phi(z, t)$ giving a time-dependent $g_{2D}(x, y, t)$.

2.2.4 Thomas-Fermi Approximation

The initial state of many experimental BEC systems is well approximated by the Thomas-Fermi wave function since most of the atoms occupy the macroscopic ground state of the trap, in which they are held, and that the kinetic energy of that state tends to be small compared to the interaction energy in the system. Setting the kinetic energy to zero in Eq. (2.8), one obtains that the Thomas-Fermi ground state wave function amplitude is given

$$|\psi(\vec{r})| = \begin{cases} \sqrt{\frac{\mu - V(\vec{r})}{g}} & \mu \geq V(\vec{r}) \\ 0 & \text{otherwise} \end{cases} \tag{2.15}$$

where μ is defined such that $\int |\psi(\vec{r})|^2 d\vec{r}^3 = N_{\text{cond}}$. In the experiments presented in this thesis, a sheet beam with narrow vertical waist provides the vertical confinement and the plane (horizontal) confinement is provided by a binary light pattern, generated by a DMD, projected onto the atom plane with sufficient strength that where the light is not masked by the DMD amplitude, there are no atoms present. The potential can thus be approximated as

$$V(\vec{r}) = V_{\text{DMD}}(x, y) + V_z(z) = V_{\text{DMD}}(x, y) + \frac{1}{2} m\omega_z^2 z^2. \tag{2.16}$$

While an approximation, it is assumed that $V_{\text{DMD}}(x, y)$ is binary, and can take on the values 0 or V_0 with $V_0 > \mu$. The chemical potential of the GPE, Eq. (2.8), in the Thomas-Fermi limit can be calculated using

$$\mu = \frac{\int \left(V(\mathbf{r}) |\Psi(\mathbf{r})|^2 + g |\Psi(\mathbf{r})|^4 \right) d^3\vec{r}}{\int |\Psi(\mathbf{r})|^2 d^3\vec{r}}. \tag{2.17}$$

In the case of the potential given in Eq. (2.16) and the Thomas-Fermi wave function, Eq. (2.15), we obtain

$$\mu = \frac{1}{2}\left(\frac{3g(m\omega_z^2)^{1/2}}{2A}\right)^{2/3} N_{cond}^{2/3} \tag{2.18}$$

where A is the 2D area of the trap where $V_{\rm DMD}(x, y) = 0$, given by $A = \int(1 - V_{\rm DMD}(x, y)/V_0)dA$.

Later, the Thomas-Fermi approximation will be used to derive an effective capacitance for atomtronic circuits. It is also used numerically as an ansatz of the ground state wave function to evolve under imaginary time before numerical simulations.

2.2.5 Healing Length

Apart from the scattering length (a_s) and the Thomas-Fermi characteristic-size of the system, the healing length ξ is another fundamental length scale relevant to the understanding of BEC physics. It is of particular importance when looking at the dynamics of defects. It is a measure of the distance over which the density makes a transition from zero density to the background level density for a bulk superfluid. It can be used as an estimate of the length scales over which features in the potential will be relevant, since the wave function will vary on length scales on the order of the healing length. It can be obtained by looking at a system of infinite extent in the y and z directions and with an infinite energy barrier for $x \leq 0$ leading to $\psi(0, y, z) = 0$ boundary condition which needs to be solved on

$$0 = \left[\frac{\hbar^2}{2m}\vec{\nabla}^2 + g|\psi(x, y, z)|^2 - \mu\right]\psi(x, y, z) \tag{2.19}$$

which is simply Eq. (2.8) without potential energy. It is also known that far from the boundary the density will 'heal' back to its bulk value, n_0, giving the second boundary condition $\lim_{x\to\infty}\psi(x, y, z) = n_0$. The lowest energy solution to this equation is one with no dependence on y or z, minimizing the kinetic energy. This makes the system easily solvable, with the solution

$$\psi(x, y, z) = \sqrt{n_0}\tanh\left(\frac{\sqrt{mn_0g}}{\hbar}x\right) = \sqrt{n_0}\tanh(x/\xi), \tag{2.20}$$

where the healing length was defined to be:

$$\xi = \frac{\hbar}{\sqrt{mn_0g}}. \tag{2.21}$$

The healing length can also be derived by looking at the minimization of the energy when transitioning from a bulk superfluid to zero density, since there is an energy cost associated with bulk density (interaction energy) and density gradients (kinetic energy) [16]. The solution leads to a healing length $\xi' = \xi/\sqrt{2}$ different by a factor of $1/\sqrt{2}$. In this thesis, Eq. (2.21) will be taken as the definition of the healing length.

2.2.6 Speed of Sound

There are many types of excitations that can occur in a condensate, with the most well-known perhaps being the topologically protected vortex and soliton excitations. Another form of excitations in the condensate are sound excitations, which are quasi-particles in the same sense as phonons in a semiconductor lattice. Their energy, in a uniform density system, is given by the Bogoliubov excitation spectrum [17]

$$\varepsilon(p) = \pm \left[\frac{gn(\vec{r})}{m} p^2 + \left(\frac{p^2}{2m} \right)^2 \right]^{1/2}, \tag{2.22}$$

where p is the momentum of the quasiparticles, which has a wavenumber and wavelength given by $k = p/\hbar$ and $\lambda = 2\pi/k$. For excitations with small momenta $p \ll 2\sqrt{mgn}$, the dispersion law is well approximated by a phonon like dispersion from

$$\varepsilon(p) \approx p\sqrt{\frac{gn(\vec{r})}{m}} = pc_s, \tag{2.23}$$

where the speed of sound is [17, 18]

$$c_s = \sqrt{\frac{gn(\vec{r})}{m}} = \sqrt{\frac{\mu}{m}} = \sqrt{n_0 g/m} = \frac{1}{\sqrt{2}} \frac{\hbar}{m\xi}. \tag{2.24}$$

This means that small amplitude and low energy (low momentum) excitations travel through the condensate the same way sound propagates through air. They do so at a constant speed know as the speed of sound c_s. The speed of sound can be used as a way to measure the chemical potential or used to check the calibration of the atom number imaging counts.

On the other side of the spectrum for large momentum $p \gg 2\sqrt{gnm}$ the dispersion relationship looks like

$$\varepsilon(p) \approx \frac{p^2}{2m} + gn \tag{2.25}$$

which is very similar to a free particle with the interaction energy acting as a potential. It is interesting that the characteristic lengths over which the excitations switch from phonon like to particle-like are when the wavenumber is the inverse healing length $k = \xi^{-1}$.

The velocity at which the energy is minimized by creating excitations can be obtained by applying the Landau criterion [19, 20]:

$$\mathscr{V}_c = \min\left(\frac{\varepsilon(p)}{p}\right) = \min\left(\frac{pc_\mathrm{s}}{p}\right) = c_\mathrm{s}. \tag{2.26}$$

The critical velocity for creation of excitations ($\mathscr{V}_c$) and the speed of sound (c_s) are one and the same.

2.2.7 *Quantum Vortices*

We can express the complex number representing the wave function in the polar representation, called Madelung transformation, as

$$\Psi(\vec{r}, t) = \sqrt{n(\vec{r}, t)}\exp\left[i\theta(\vec{r}, t)\right], \tag{2.27}$$

where the $n(\vec{r}, t)$, $\theta(\vec{r}, t)$ represent the density and phase of the BEC, respectively. This is equivalent to treating the BEC as tiny droplets of superfluid with a certain density and velocity dictated by the phase through the probability current

$$\vec{j}(\vec{r}, t) = \frac{\hbar}{2mi}\left[\Psi^*(\vec{r}, t)\vec{\nabla}\Psi(\vec{r}, t) - \Psi(\vec{r}, t)\vec{\nabla}\Psi^*(\vec{r}, t)\right]. \tag{2.28}$$

Substituting in the hydrodynamic form of the wave function, Eq. (2.27), the probability current becomes

$$\vec{j}(\vec{r}, t) = \frac{\hbar}{m}n(\vec{r}, t)\vec{\nabla}\theta(\vec{r}, t) \tag{2.29}$$

which, as expected of a current, is a density multiplied by the velocity

$$\vec{v}(\vec{r}, t) = \frac{\vec{j}(\vec{r}, t)}{n(\vec{r}, t)} = \frac{\hbar}{m}\vec{\nabla}\theta(\vec{r}, t). \tag{2.30}$$

The phase of the wave function can be thought of as a potential giving rise to the velocity field in the condensate, with the superfluid flowing along its gradient.

A consequence of Eq. (2.30) for a single-valued phase is that the vorticity $\vec{\omega}$ is 0 everywhere:

$$\vec{\omega}(\vec{r}, t) \equiv \vec{\nabla} \times \vec{v}(\vec{r}, t) = \frac{\hbar}{m}\vec{\nabla} \times \vec{\nabla}\theta(\vec{r}, t) = 0, \tag{2.31}$$

leading to an irrotational velocity field. To support circulation, a pole in the phase profile is required which is called a vortex. This pole must lead to a single-valued wave function, therefore the circulation Γ around any closed surface contour $\mathscr{C}$ meaning

$$\Gamma \equiv \oint_{\mathscr{C}} \vec{v}(\vec{r},t)\cdot d\vec{r} = \frac{\hbar}{m}\oint_{\mathscr{C}} \vec{\nabla}\theta(\vec{r},t)\cdot d\vec{r} = \frac{\hbar}{m}2N\pi = \kappa N \tag{2.32}$$

with $N \in \mathbb{Z}$ and $\kappa = h/m$ also known as the quantum circulation. This means that circulation in the condensate is quantized to be an integer multiple of κ. The consequence is that, contrary to vortices in a classical fluid, the vortices in the superfluid must have quantized circulation since any path travelled through the condensate must undergo a phase change which is a multiple of 2π as a consequence of the wave function being single-valued. The value of N determines the winding number, or charge, of the vortex by representing how many times the phase went through a 2π winding. The winding number of a single vortex is usually ± 1 due to the high energy cost of higher charged vortices which tend to decay into multiple singly-charged vortices [21, 22]. This leads to vortices with a clockwise and counter-clockwise circulation which are often referred to as antivortices and vortices, since combining them results in annihilation of the flow. At the singularity, at the vortex location, the wave function vanishes to prevent infinite kinetic energy and to keep the wave function single-valued.

Assuming a vortex is located a position $\vec{r}_0$ the vortex circulation, Eq. (2.32), can be calculated using the Stokes' theorem assuming the contour $\mathscr{C}$ encloses only one vortex core with joining surface $\mathscr{S}$ confined to the superfluid:

$$\oint_{\mathscr{C}} \vec{v}(\vec{r},t)\cdot d\vec{r} = \int_{\mathscr{S}} \vec{\nabla}\times\vec{v}(\vec{r},t)\cdot d\vec{S} = \kappa N. \tag{2.33}$$

Assuming we are working with 2D vortices in the xy-plane this results in the vorticity of vortices being given by

$$\vec{\omega}(\vec{r},t) = \kappa N \hat{z}\delta\left(|\vec{r}-\vec{r}_0|\right). \tag{2.34}$$

The singularity leads to the vorticity being confined to the vortex core location $\vec{r}$. These planar vortices in a quasi two-dimensional condensate are thus *point vortices*, with only a minor deviation being a correction due to the density envelope.

The velocity field produced by a vortex is obtained by minimizing the energy it produces. This leads to a phase profile which is azimuthally symmetric [20]. The field needs to satisfy Eq. (2.32) leading to

$$\oint_{\mathscr{C}} \vec{v}_v(\vec{r},t)\cdot d\vec{r} = v_v\hat{\phi}\cdot 2\pi r\hat{\phi} = \kappa N \tag{2.35}$$

where $\hat{\phi}$ is the polar coordinate unit vector for a coordinate system centred on the vortex core. By re-arranging, the velocity field produced by the vortex is

$$\vec{v}_v(\vec{r}) = \frac{\hbar}{m}\frac{N}{|\vec{r}-\vec{r}_0|}\hat{\phi}. \tag{2.36}$$

A diverging velocity can be seen for $\vec{r} \to \vec{r}_0$, leading to diverging of the kinetic energy, but this is prevented by the density envelope of the vortex [9]

$$n_v(\vec{r}) = \frac{n_0|\vec{r}-\vec{r}_0|^2}{|\vec{r}-\vec{r}_0|^2 + 2\xi^2} \tag{2.37}$$

where n_0 is the background density and $n_v(\vec{r})$ is an approximation analytical form of the density profile that minimizes the energy of the vortex. Close to the core, the density tends to zero and far from the core the density stabilizes to the background density. Density profile solutions for higher charge cores show that they are unstable and will tend to decay into the appropriate number of single winding vortices [21, 22], unless the external potential is modified in such a way as to stabilize them [22]. The wave function for an unbounded infinite uniform density superfluid system with a stable single-charge vortex situated at the origin is

$$\Psi(\vec{r}, t) = n_v(\vec{r})e^{iN\phi} \tag{2.38}$$

with $N \in \{-1, 1\}$.

Since the vorticity, Eq. (2.34), and velocity profile, Eq. (2.36), of a quantum vortex strongly resemble that of a point vortex, under the right conditions quantum vortices should behave like point vortices. Assuming that the intervortex spacing is well above the healing length $l \gg \xi$ and that the vortex bending is suppressed by having the trap frequency ratio exceed $\omega_z : \omega_r = 8 : 1$ [23]. This is typically the case for quasi two-dimensional systems working with highly oblate potentials like ours. It was shown that to first order the dynamics of the quantum vortices reduces down to the point vortex model [24–31]. This reduction results in the following equation for the velocity of a vortex, advected by the flow:

$$\dot{\vec{r}}_j = \frac{\hbar}{m}\sum_{i\neq j}\vec{\nabla}\theta_i|_{\vec{r}=\vec{r}_j}. \tag{2.39}$$

The velocity of the vortex j is thus due to the local phase gradient imprinted by all the other vortices in the system.

2.2.8 *Quantum Hydrodynamics*

In this thesis, we will be looking at the analogues between the superfluids, acoustic circuits and electronic circuits. To help with the comparison between the different systems and to have a deeper understanding of the transport dynamics, the hydrodynamics form of the Gross-Pitaevskii equation is especially useful and reveals a close resemblance with the motion of sound in an acoustic system which is the subject of Chap. 5.

Substituting the Madelung transformation [Eq. (2.27)] into the Gross-Pitaevskii equation [Eq. (2.3)] allows for the expression of the GPE in its hydrodynamic form. By separating the real and imaginary components, one obtains a system of coupled equations

$$m\frac{\partial \vec{v}(\vec{r},t)}{\partial t} = -\vec{\nabla}\left[\frac{1}{2}m\vec{v}(\vec{r},t) + V(\vec{r},t) + gn(\vec{r},t) - \frac{\hbar^2}{2m\sqrt{n(\vec{r},t)}}\vec{\nabla}^2\vec{v}(\vec{r},t)\right], \tag{2.40a}$$

$$\frac{\partial n(\vec{r},t)}{\partial t} + \vec{\nabla}\cdot(n(\vec{r},t)\vec{v}(\vec{r},t)) = 0. \tag{2.40b}$$

The mass continuity equation, Eq. (2.40b), was obtained by matching the imaginary terms, and ensuring conservation of the number of particles in the system. The real part, represented in Eq. (2.40a), is an equivalent to the Euler equation of fluid dynamics, but for a quantum system. It resembles Newton's equation, with an effective potential given by the terms inside the bracket. These equations of motions are very close to their classical equivalents, except the quantum one is irrotational, $\vec{\nabla}\times\vec{v}(\vec{r},t) = 0$, in the absence of vortices, and the quantum equation contains an extra term, the *quantum pressure* (rightmost term in Eq. (2.40a)), which results from the compressibility of quantum fluids.

The resulting quantum Euler equation system energy can be broken down into its various components:

$$E_H(t) = \frac{m}{2}\int n(\vec{r},t)|\vec{v}(\vec{r},t)|^2 d^3\vec{r}, \tag{2.41a}$$

$$E_Q(t) = \frac{\hbar^2}{2m}\int |\vec{\nabla}\sqrt{n(\vec{r},t)}|^2 d^3\vec{r}, \tag{2.41b}$$

$$E_K(t) = E_H(t) + E_Q(t), \tag{2.41c}$$

$$E_V(t) = \int n(\vec{r},t)V(\vec{r},t)d^3\vec{r}, \tag{2.41d}$$

$$E_I(t) = \frac{g}{2}\int n(\vec{r},t)^2 d^3\vec{r}. \tag{2.41e}$$

They represent the hydrodynamic kinetic energy, quantum pressure energy, total kinetic energy, potential energy, and interaction energy, respectively. It is possible to use the Helmholtz decomposition to separate the hydrodynamics energy further into its incompressible and compressible components [32]. For a uniform density condensate, this can be done achieved by decomposing the velocity field into

$$\vec{v}(\vec{r},t) = \vec{v}_{\mathrm{In}}(\vec{r},t) + \vec{v}_{\mathrm{C}}(\vec{r},t), \tag{2.42}$$

where $\vec{\nabla} \cdot \vec{v}_{\mathrm{In}}(\vec{r}, t) = 0$ and $\vec{\nabla} \times \vec{v}_{\mathrm{C}}(\vec{r}, t) = 0$. The incompressible velocity component, $\vec{v}_{\mathrm{In}}(\vec{r}, t)$, contains the motion due to the vortices in the system, while the compressible component, $\vec{v}_{\mathrm{C}}(\vec{r}, t)$, relates to the motion induced by the phonons in the system. The decomposition of the hydrodynamic kinetic energy can be expressed as

$$E_{\mathrm{In}}(t) = \frac{m}{2} \int n(\vec{r}, t) |\vec{v}_{\mathrm{In}}(\vec{r}, t)|^2 d^3\vec{r}, \tag{2.43a}$$

$$E_{\mathrm{C}}(t) = \frac{m}{2} \int n(\vec{r}, t) |\vec{v}_{\mathrm{C}}(\vec{r}, t)|^2 d^3\vec{r}, \tag{2.43b}$$

which are the incompressible and compressible contribution to the hydrodynamic quantum energy, respectively. The kinetic energy can therefore be expressed as $E_{\mathrm{K}}(t) = E_{\mathrm{Q}}(t) + E_{\mathrm{In}}(t) + E_{\mathrm{C}}(t)$.

2.3 Point Vortex Model

Quasi two-dimensional quantum fluid systems have the advantage that the velocity field created by the vortices is constrained to a plane, due to the suppression of vortex bending which leads to all the vortices being parallel to the plane normal. In the limit where the density is uniform, the vorticity of the field is well localized, and the vortical particles are well separated. The vorticity field can be well approximated by a field of localized point-particle elements of rotational flow (point vortices). The vorticity field of a point vortex system can be expressed as

$$\vec{\omega}(\vec{r}) = \sum_{i}^{N_v} \Gamma_i \delta\left(\vec{r} - \vec{r}_i\right) \hat{z}, \tag{2.44}$$

where the circulations of the point vortices are given by $\{\Gamma_i\}$ and their position by $\{\vec{r}_i\}$ with $i \in \{1, 2, ..., N_v\}$. By the same energy minimization argument as in Sect. 2.2.7, the velocity field of a point vortex is $\vec{v}_i = \Gamma_i / 2\pi |\vec{r} - \vec{r}_i| \hat{\phi}_i$ where $\hat{\phi}_i$ is the tangential unit vector of the vortex at $\vec{r}$. The velocity field for this vortex distribution is then given by

$$\vec{v}(\vec{r}) = \sum_{i}^{N_v} \frac{\Gamma_i}{2\pi |\vec{r} - \vec{r}_i|} \hat{\phi}_i \tag{2.45}$$

This leads to a system of coupled equations of motion for the point vortices [33, 34]

$$\frac{d\vec{r}_l}{dt} = \frac{1}{2\pi} \sum_{i \neq l} \Gamma_i \frac{\vec{r}_l - \vec{r}_i}{|\vec{r}_l - \vec{r}_i|^2} \times \hat{z}, \tag{2.46}$$

where the motion of a point-vortex is dictated by the local flow generated by the other vortices exactly the same situation as in Eq. (2.39). The kinetic energy contained in the system can be computed from the generated flow field [Eq. (2.45)] using

$$H = \frac{1}{2}\int \rho(\vec{r})|\vec{v}(\vec{r})|^2 d^2\vec{r}. \tag{2.47}$$

The kinetic energy of an isolated 2D fluid containing N_v point vortices can be expressed in terms of the relative vortex positions [35]. In an unbounded uniform fluid, it has the form

$$H = -\frac{\rho_0}{4\pi}\sum_{i\neq j}\Gamma_i\Gamma_j \ln|\vec{r}_i - \vec{r}_j|, \tag{2.48}$$

where $\rho_0 = mn_0$ is the background fluid density. The equations of motion can be expressed in their canonical form using standard Hamilton's equations

$$\Gamma_i\frac{dx_i}{dt} = \frac{\partial H}{\partial y_i}, \qquad \Gamma_i\frac{dy_i}{dt} = -\frac{\partial H}{\partial x_i}. \tag{2.49}$$

From Eq. (2.48), it might seem as though all the energy of the point vortices is due to interaction energy. In reality, the energy is all kinetic energy from the motion of the fluid induced by the presence of the vortical particles. Point vortices are 'strange particles' in the sense that there is no mass associated with them and their motion does not follow a Newtonian-like equation but instead follow a first order differential equation. The interaction potential of point vortices is very reminiscent of logarithmic scaling of the interaction between standard electric charges, familiar methods used to solve the electric potential with given boundary conditions should also apply to point vortices. In Eqs. (2.48) and (2.49), the canonical conjugate variables are positions (x, y) instead of position and momentum (x, P_x) which means that the phase space of the system is fully determined by the position of the vortices, leading to profound changes in statistical properties of the particles compared with standard particles. In an open unbounded system, there are two quantities that are conserved under evolution under Eq. (2.46):

$$\vec{P} = \sum_i^{N_v}\Gamma_i\vec{r}_i, \tag{2.50a}$$

$$\vec{L} = \sum_i^{N_v}\Gamma_i|\vec{r}_i|^2\hat{z}. \tag{2.50b}$$

The linear momentum, $\vec{P}$, and the angular momentum, $\vec{L}$, are conserved due to the translation and rotational symmetry of an infinite system.

2.3.1 Circular Domain

While unbounded system are interesting from a theoretical point of view, experimental systems are bounded. A simple geometry is the circularly bounded domain of radius R. This geometry requires that on the domain $\partial\mathscr{D}$ of the disc, the flow normal to the surface be zero:

$$\vec{v}\cdot\hat{n}\Big|_{\partial\mathscr{D}} = 0. \tag{2.51}$$

Since point vortices behave like charges, this is the same as solving the Laplace equation on a circular domain of constant potential which can be done through the method of images [36], which consists of placing 'imaginary' images of the charges outside the domain so as to satisfy the boundary condition. To satisfy the boundary condition, an image vortex of opposite sign but equal circulation $\bar{\Gamma}_i = -\Gamma_i$ is placed at the location

$$\bar{\vec{r}}_i = \frac{R^2\vec{r}_i}{|\vec{r}_i|^2} \tag{2.52}$$

which is outside the boundary. The Hamiltonian for point vortices inside a circular domain becomes [34, 35, 37–40]

$$H_0 = \frac{\rho_0}{4\pi}\left[\sum_{i,j}\Gamma_i\Gamma_j \ln\left|\frac{\vec{r}_i - \bar{\vec{r}}_j}{R}\right| - \sum_{i\neq j}\Gamma_i\Gamma_j \ln\left|\frac{\vec{r}_i - \vec{r}_j}{R}\right|\right], \tag{2.53}$$

which is Eq. (2.48) but bonded with the extra interaction due to the image vortices. The equations of motion on the circular domain are

$$\frac{d\vec{r}_l}{dt} = \frac{1}{2\pi}\left[\sum_{i\neq l}\Gamma_i \frac{\vec{r}_l - \vec{r}_i}{|\vec{r}_l - \vec{r}_i|^2} \times \hat{z} + \sum_i \bar{\Gamma}_i \frac{\vec{r}_l - \bar{\vec{r}}_i}{|\vec{r}_l - \bar{\vec{r}}_i|^2} \times \hat{z}\right]. \tag{2.54}$$

By Noether's theorem, only the angular momentum is conserved on the circular domain.

2.3.2 Elliptical Domain

Beyond simple circular geometries the point vortex model can be extended for any simply connected domain. This is achieved by using a conformal map to the unit disc combined with the method of images. Under a conformal map $\zeta = f(z)$, which derives the vortex motion in the domain $z \in \Omega$ from that in the domain $\zeta \in \mathscr{D}$, the Hamiltonians are related via [34]

$$H_\Omega(z_1, \ldots, z_N) = H_\mathscr{D}(\zeta_1, \ldots, \zeta_N) - \frac{\rho_0}{4\pi} \sum_{j=1}^{N_v} \Gamma_i^2 \ln \left| \frac{d\zeta}{dz} \right|_{z=z_i}, \tag{2.55}$$

where $z_i = x_i + iy_i$. If the map $\zeta = f(z)$ transforms a (simply connected) domain Ω to the unit disk $\mathscr{D} = \{\zeta \in \mathbb{C} \mid |\zeta| \leq 1\}$, Eq. (2.55) gives [34]

$$H_\Omega = -\frac{\rho_0}{4\pi} \left[\sum_i \Gamma_i^2 \ln \left| \frac{\zeta'(z_i)}{1-|\zeta_i|^2} \right| + \sum_{i,k}' \Gamma_i \Gamma_k \ln \left| \frac{\zeta_i - \zeta_k}{1-\zeta_i \zeta_k^*} \right| \right], \tag{2.56}$$

where $\zeta_j \equiv f(z_j)$ and the prime on the second sum indicates the exclusion of the term $j = k$. The domain of the ellipse interior, $\Omega = \left\{ z \in \mathbb{C} \mid \Re(z)^2/a^2 + \Im(z)^2/b^2 \leq 1 \right\}$ is mapped to the unit disk by the conformal map [41]

$$\zeta = f(z) = \sqrt{k} \, \mathrm{sn} \left(\frac{2K(k)}{\pi} \sin^{-1} \left(\frac{z}{\sqrt{a^2 - b^2}} \right) ; k \right). \tag{2.57}$$

Here $\mathrm{sn}\,(z\,; k)$ is the Jacobi elliptic sine function, $K(k)$ is the complete elliptic integral of the first kind, and k is the elliptical modulus, given by

$$k^2 = 16\rho \prod_{n=1}^{\infty} \left(\frac{1 + \rho^{2n}}{1 + \rho^{2n-1}} \right)^8, \tag{2.58}$$

where $\rho = (a - b)^2/(a + b)^2$. The equations of motion for the vortices in the Ω domain can be calculated from the $\mathscr{D}$ domain using Routh's rule [34] which gives the following

$$v_x^\Omega(z_l) - iv_y^\Omega(z_l) = \left[v_x^\mathscr{D}(\zeta_l) - iv_y^\mathscr{D}(\zeta_l) \right] \cdot \left. \frac{d\zeta}{dz} \right|_{z=z_l} - \frac{i\Gamma_l}{4\pi} \cdot \left. \frac{f''(z)}{f'(z)} \right|_{z=z_l}, \tag{2.59}$$

where z_l is the position of the point-vortex. The velocity field in the circular domain is a known quantity which allows for the evolution of the system in the elliptical domain by simply having to compute the derivatives of the conformal map at the vortex locations. The rightmost term simply removes the contribution of the vortex itself to the local velocity while still accounting for velocity due to the boundary condition. Another quantity of interest is the velocity field in the original domain away from the point-vortices which can simply be calculated using

$$v_x^\Omega(z) - iv_y^\Omega(z) = \left[v_x^\mathscr{D}(\zeta) - iv_y^\mathscr{D}(\zeta) \right] \cdot \frac{d\zeta}{dz} \tag{2.60}$$

which is a simplified version of Eq. (2.59) since everywhere there is no vortex, the vorticity of the field is zero.

2.4 Manipulating Atoms with Light

2.4.1 Conservative Optical Dipole Potentials

Light fields can be used to trap neutral atoms through the interactions of the AC electric field with the internal states of the electrons making up the atoms by inducing a dipole moment. It leads to a dipole interaction of the form:

$$U_{\text{dipole}}(\vec{r}, t) = \gamma I_{\text{light}}(\vec{r}, t). \tag{2.61}$$

For a two-level system, if the rotating wave approximation holds, then the conversion from light field to dipole interaction and scattering rate are given by [42]:

$$\gamma = \frac{3\pi c^2}{2\omega_0^3}\frac{\Gamma}{\Delta}, \tag{2.62a}$$

$$\Gamma_{\text{scatter}}(\vec{r}, t) = \frac{3\pi c^2}{2\hbar\omega_0^3}\left(\frac{\Gamma}{\Delta}\right)^2 I_{\text{light}}(\vec{r}, t). \tag{2.62b}$$

Where Γ is the decay rate of the transitions, Δ is the detuning from the two-level transition, ω_0 is the transition frequency, and c is the speed of light. In the case of $\Delta < 0$ the potential is minimum where the light intensity is highest which is commonly refereed to as red detuning. Conversely, for $\Delta > 0$ the potential is known as blue detuned and the atoms are repelled by the light. From Eqs. (2.61) and (2.62a), it can be seen that the dipole force scales as I/Δ whereas from Eq. (2.62b) the scattering scales as I/Δ^2. This means that to prevent heating due to scattering far-off-resonant light with high laser powers are preferred.

The conversion from intensity to potential energy for a multiple level system is due to the dipole interaction which is given by [42]

$$\gamma = \frac{\pi c^2 \Gamma}{2\omega_0^3}\left(\frac{2 + \mathcal{P} g_F m_F}{\Delta_{2,F}} + \frac{1 - \mathcal{P} g_F m_F}{\Delta_{1,F}}\right) \tag{2.63}$$

where $\mathcal{P}$ characterizes laser polarization (0 for linear, ± 1 for circularly polarized $\sigma^{\pm}$), g_F is the Landé factor, m_F is the magnetic hyperfine structure level, and $\Delta_{2,F}(\Delta_{1,F})$ is the detuning from the ${}^2S_{1/2} \rightarrow {}^2P_{3/2}({}^2S_{1/2} \rightarrow {}^2P_{1/2})$ transitions for the alkali atoms. In our experiment, we use ^{87}Rb with transitions shown in Fig. 2.2.

2.4.2 Phase Imprinting

The quickly varying potentials used in time averaging tend to modify the local phase through phase imprinting. This effect will be used in Chap. 4 to create a quantized current in a ring. In this section, we simply derive the general equation for phase imprinting.

If one assumes that the potential changes quickly enough that the density is time-independent, and can be approximated by the density due to the time-averaged effective static trap, then the wave function can be approximated as $\Psi(\vec{r},t) = A(\vec{r})\exp\left[-i\phi(\vec{r},t)\right]$. Substituting this wave function into the GPE Eq. (2.3), we get

$$\hbar\frac{\partial\phi(\vec{r},t)}{\partial t} = -\frac{\hbar^2}{2m}\left(\nabla\phi(\vec{r},t)\cdot\nabla\phi(\vec{r},t) + \frac{\nabla^2 A(\vec{r})}{A(\vec{r})} - i\left[\nabla^2\phi(\vec{r},t) + 2\frac{\nabla A(\vec{r})\cdot\nabla\phi(\vec{r},t)}{A(\vec{r})}\right]\right) + V(\vec{r},t) + gA^2(\vec{r}). \tag{2.64}$$

In the limit of short imprinting time, such that there is no motion of the atoms, and ignoring the constant terms, Eq. (2.64) reduces to the phase imprinting equation [43]:

$$\frac{\partial\phi(\vec{r},t)}{\partial t} = \hbar^{-1}V(\vec{r},t). \tag{2.65}$$

The imprinted phase is thus due to the instantaneous potential $V(\vec{r},t)$.

2.5 Data Acquisition

In this section, two ways in which the density distribution of the cloud is determined are described. These consist of absorption imaging and Faraday imaging which are scattering and phase based, respectively. A basic overview of the light-atom interaction involved in both processes, and the main parameters of interest during the imaging process are detailed. To achieve an accurate atom count, a calibration of the images was performed and the calibration technique is presented. The basics properties of Bragg-scattering are also presented, since it is used to experimentally probe the momentum spectrum of the BECs.

Absorption imaging is used for imaging along both the horizontal ($\hat{x}$) and vertical ($\hat{z}$) directions, whereas Faraday imaging is only along the high-resolution vertical direction ($\hat{z}$). These are used together to directly measure atomic population in reservoirs (Chap. 5) and measure vortex positions (Chap. 6). Combining absorption imaging with Bragg-scattering (Sect. 2.5.4) allows the measurement of the sign of the vortices (Sect. 6.5) by probing the momentum distribution of the cloud. Combining vertical images with time of flight gives information about the condensate fraction, temperature, and momentum distribution of the atomic cloud.

2.5.1 Absorption Imaging

In cold atoms, the most common and simple way to image a cloud is through the use of absorption imaging, which works by measuring the attenuation of the light transmitted through the cloud due to scattering. The procedure consists of shining a beam of resonant light through the atoms, and an imaging system is constructed to direct the light onto a camera (CCD or EMCCD). Three images are taken: one of the light in the presence of the atoms (Fig. 2.1B), one without atoms (Fig. 2.1A) and one without light to characterise systematic noise. Through scattering of some of the light by the atoms, a shadow is recorded by the camera, which, when contrasted with the light without atoms, can be used to extract the atom spatial distribution and number. Throughout the experiments described in this thesis, absorption imaging along the horizontal ($\hat{x}$) axis is used to extract the atom number and condensate fraction. Along the vertical ($\hat{z}$) axis, absorption imaging combined with Bragg-spectroscopy is used to measure the velocity distribution.

Imaging is performed on the D_2 line of ^{87}Rb on the and $5^2\mathrm{S}_{1/2}|\mathrm{F}=1, m_\mathrm{F}=-1\rangle \rightarrow 5^2\mathrm{P}_{3/2}|\mathrm{F}'=3\rangle$ transition (Fig. 2.2). Since the BEC is initially prepared in the $5^2\mathrm{S}_{1/2}|\mathrm{F}=1, m_\mathrm{F}=-1\rangle$ state, it is first transferred to the $5^2\mathrm{S}_{1/2}|\mathrm{F}=2\rangle$ state through a repump pulse to the $5^2\mathrm{P}_{3/2}|\mathrm{F}'=2\rangle$ which can then spontaneously decay to the desired $5^2\mathrm{S}_{1/2}|\mathrm{F}=2\rangle$ state, or back down to the $5^2\mathrm{S}_{1/2}|\mathrm{F}=1\rangle$. Even though two decay channels are possible, through many cycles most of the atoms will end up in the

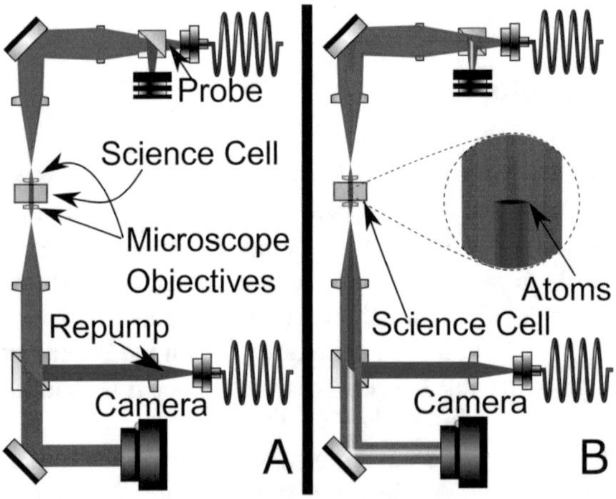

Fig. 2.1 Physical implementation of absorption imaging in our system. **A** shows the imaging system without the presence of atoms, the repump laser ($5^2\mathrm{S}_{1/2}|\mathrm{F}=1\rangle \rightarrow 5^2\mathrm{P}_{3/2}|\mathrm{F}'=2\rangle$) is depicted in blue, the probing light ($5^2\mathrm{S}_{1/2}|\mathrm{F}=2\rangle \rightarrow 5^2\mathrm{P}_{3/2}|\mathrm{F}'=3\rangle$) in red and the overlap of the two lasers is shown in pink. This configuration is used to image the light without the atoms. **B** shows the imaging with the cloud present. As can be seen from the inset, the repump laser is absorbed by the atoms that fell out of the cyclical transition into $5^2\mathrm{S}_{1/2}|\mathrm{F}=1\rangle$ to put them back into $5^2\mathrm{S}_{1/2}|\mathrm{F}=2\rangle$. Then the probing light is absorbed and scattered by the atoms leaving a visible spatially variable attenuation of the light detected at the camera, which can be contrasted to light profile than using **A** to obtain the spatial profile of the atoms integrated along the propagation direction of the probing beam

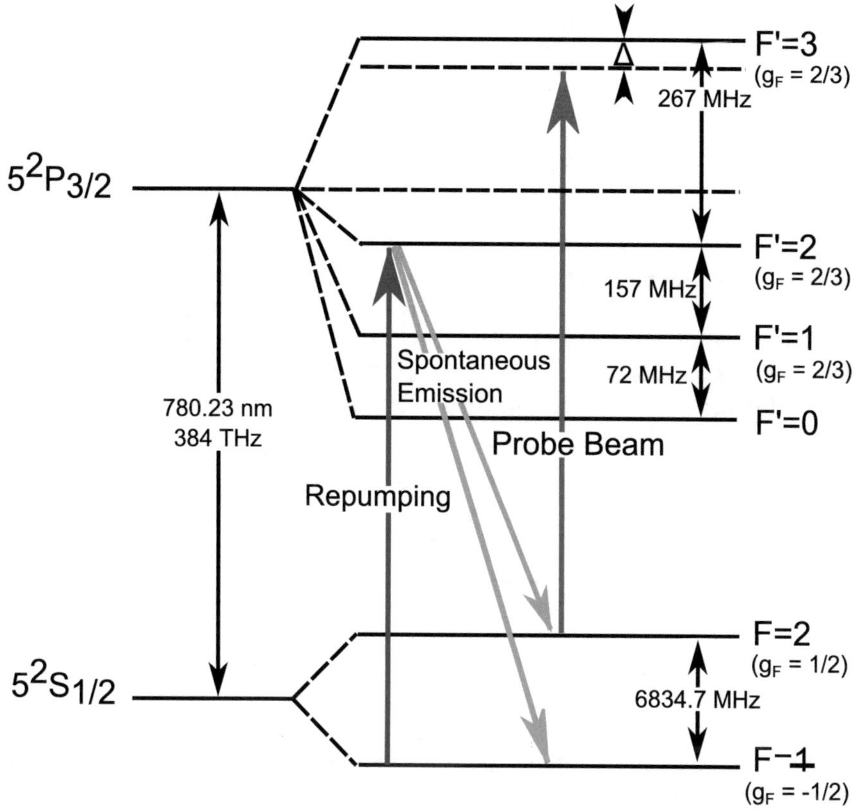

Fig. 2.2 Atomic-level diagram for ^{87}Rb on the D2 line with corresponding Landé g_F-factors for each of the hyperfine levels splitting between the levels [46]. Depicted is the optical scheme for the absorption imaging where pre-imaging the atoms are optically pumped (blue arrow) from $5^2S_{1/2}|F = 1\rangle \rightarrow 5^2P_{3/2}|F' = 2\rangle$, followed by a spontaneous emission (orange arrow) back down to either $5^2S_{1/2}|F = 1\rangle$ or $5^2S_{1/2}|F = 1\rangle$. This causes accumulation of atom in the $5^2S_{1/2}|F = 2\rangle$ dark state over many cycle. During the imaging step, a probe laser (red arrow) with detuning Δ illuminates the atoms, transmitting them from the $5^2S_{1/2}|F = 2\rangle$ state to the $5^2P_{3/2}|F' = 3\rangle$ state

$5^2S_{1/2}|F = 2\rangle$ state. The imaging light probes the $5^2S_{1/2}|F = 2\rangle \rightarrow 5^2P_{3/2}|F' = 3\rangle$ transition and is detuned by Δ. Linearly polarized light is used assuming a uniform distribution of atoms in each of the $5^2S_{1/2}|F = 2\rangle$ sub-states due to the degeneracy of the magnetic states in small magnetic fields. Another way of driving the $5^2S_{1/2}|F = 1\rangle \rightarrow 5^2S_{1/2}|F = 2\rangle$ is directly through the microwave transition of 6.8 GHz [44, 45]. This has the advantage of being more efficient, since it does not rely on multiple cycle and fractional transfer to accumulate atoms into the $5^2S_{1/2}|F = 2\rangle$ state. The optical method has the advantage of requiring no additional experimental equipment.

The attenuation of the probing light travelling through the condensate is given by the Beer-Lambert law. Accounting for the effects of the detuning of the probe beam and the possible saturation in the cross section [47], the attenuation of the probe light intensity while passing through the atomic cloud can be written as:

$$\frac{dI(\vec{r})}{d\hat{k}} = -n(\vec{r})\sigma(\vec{r})I(\vec{r}) = -n(\vec{r})\sigma_0\left(\frac{I(\vec{r})}{1+4(\Delta/\Gamma)^2 + I(\vec{r})/I_{\text{sat}}}\right) \tag{2.66}$$

$I(\vec{r})$ is the incident probe intensity, $\hat{k}$ is the propagation direction of the probe beam, $\sigma(\vec{r})$ is the atomic cross-section, σ_0 is the two-level atom resonant absorption cross-section, $n(\vec{r})$ is the local 3D density, I_{sat} is the saturation intensity of the probed two-level transition, Γ is the two-level transition linewidth and $\Delta = \omega - \omega_0$ is the detuning form the resonance, where ω_0 (ω) are the resonance transition (probe beam) angular frequency. The on resonance two level cross-section can be calculated by [46, 48]

$$\sigma_0 = \frac{\hbar\omega_0\Gamma}{2I_{\text{sat}}} \tag{2.67}$$

and the saturation can be calculated using

$$I_{\text{sat}} = \frac{c\varepsilon_0\Gamma^2\hbar^2}{4\left|\hat{\varepsilon}\cdot\vec{d}\right|^2}, \tag{2.68}$$

where ε_0 is the permittivity of free space, $\hat{\varepsilon}$ is the unit polarization vector of the probe beam electric field, $\vec{d}$ is the atomic dipole moment and c is the speed of light. In the case of linearly polarized probe beam tuned to the $5^2\text{S}_{1/2}|\text{F}=2\rangle \rightarrow 5^2\text{P}_{3/2}|\text{F}'=3\rangle$ transition, and atoms are uniformly distributed in the $5^2\text{S}_{1/2}|\text{F}=2\rangle$ state, the saturation intensity and the cross section are calculated to be $I_{\text{sat}} = 3.42\ \text{mW/cm}^2$, $\sigma_0 = 7\lambda^2/10\pi = 1.4\times10^{-13}\ \text{m}^{-2}$.

Assuming that the probe beam is travelling in the $\hat{x}$ direction, the column density $n_{2\text{D}}(y,z) = \int n(\vec{r})dx$ along the beam propagation is obtained by integrating Eq. (2.66) giving

$$n_{2\text{D}}(y,z) = \frac{1}{\sigma_0}\left(\left[1+4\left(\frac{\Delta}{\Gamma}\right)^2\right]\ln\left[\frac{I_{\text{P}}(y,z)}{I_{\text{M}}(y,z)}\right] + \frac{I_{\text{P}}(y,z)-I_{\text{M}}(y,z)}{I_{\text{sat}}}\right), \tag{2.69}$$

where $I_{\text{P}}(y,z)$ is the beam intensity of the probe beam and $I_{\text{M}}(y,z)$ is the measured intensity at the camera after the probe has interacted with the atom cloud. For imaging intensities much lower than the saturation intensity ($1+4(\Delta/\Gamma)^2 \gg I(\vec{r})/I_{\text{sat}}$), we can ignore the second term and we obtain that

$$n_{2\text{D}}(y,z) \approx \frac{1}{\sigma_0}\left[1+4\left(\frac{\Delta}{\Gamma}\right)^2\right]\ln\left[\frac{I_{\text{P}}(y,z)}{I_{\text{M}}(y,z)}\right]. \tag{2.70}$$

To obtain $I_P(y, x)$ and $I_M(y, z)$, we take an image $I_w(y, z)$ with the atoms and probe beam on (Fig. 2.1B), $I_{wo}(y, z)$ without the atoms and probe on (Fig. 2.1A), and $I_{dark}(y, z)$ without atoms and probe off. Then $I_P(y, z) = I_{wo}(y, z) - I_{dark}(y, z)$ and $I_M(y, z) = I_w(y, z) - I_{dark}(y, z)$.

In most experiments some saturation will occur but it is impractical to correct for the saturation using Eq. (2.69) due to the dependence on the probe beam intensity which would fluctuate for any variation in the probe beam power. Instead, it is useful to make the approximation that there exists an effective intensity (I_{eff}) spatially invariant going through the cloud that corrects for the saturation. In this case the solution to Eq. (2.66) becomes

$$n_{2D}(y, z) \approx \frac{1}{\sigma_0}\left[1 + 4\left(\frac{\Delta}{\Gamma}\right)^2 + \frac{I_{eff}}{I_{sat}}\right] \ln\left[\frac{I_P(y, z)}{I_M(y, z)}\right]. \tag{2.71}$$

Inspired by [49], to find I_{eff}, the integrated optical density of our system is repeatedly measured

$$OD_{int} = \int dz dy \ln\left[\frac{I_P(y, z)}{I_M(y, z)}\right] = \frac{\sigma_0 N}{1 + 4\left(\frac{\Delta}{\Gamma}\right)^2 + \frac{I_{eff}}{I_{sat}}} \tag{2.72}$$

over many detunings for a fixed atom number cloud shown in Fig. 2.3A. For any Δ the total number of atoms ($N = \int n_{2D}(y, z) dy dz$) and the effective intensity are fixed, but unknown in Eq. (2.72), they will be left as free parameters which are fitted to the obtained optical densities (ODs). The best-fit for our system was found to be $I_{eff} = (2.1 \pm 0.1) \times I_{sat}$ mostly due to the probe beam focusing about 50 mm from the atom cloud. Using Eq. (2.71) the atom number in the system as a function of detuning is calculated and shown in Fig. 2.3B, and it is a constant as expected. The validity of the obtained atom number was also independently verified by measuring the speed of sound in the system, and by measuring the chemical potential through the density-dependent shift of the Bragg resonance.

2.5.2 *Bimodal Fitting*

Extracting useful information from the absorption imaging, other than the total atom number, requires fitting the cloud to a number of parameters. The parameters of interest are the dimensions of the thermal cloud and those of the condensate. These fits allow the extraction of the respective atom number in the condensate and in the thermal cloud, allowing calculation of the condensate fraction. To extract the chemical potential of the condensate, the evolution of the waist of the condensate as a function of time during free expansion is monitored. Similarly, the temperature can be obtained by doing the same for TOF expansion of the waist of the thermal gas. A

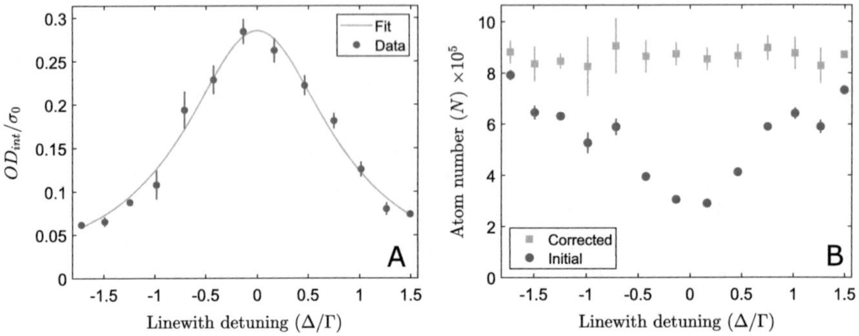

Fig. 2.3 Calibration of the absorption imaging. **A** Plot of the integrated optical density versus detuning of the probe laser, for images of an atomic cloud with fixed atom number. The fit is a two parameters fit to Eq. (2.72) with I_{eff} and N as free parameters. Where I_{eff} is the effective saturation correction factor. The detuning is normalized by transition linewidth $\Gamma = 2\pi \times 6.07$ MHz. **B** Shows the atom number as calculated with and without I_{eff} correction for saturation. As can be seen, once the correction is taken into account the calculated atom number is constant regardless of detuning

Gaussian is used to fit our thermal cloud since it is the final shape of any cloud with an Boltzmann momentum distribution in the limit of long free expansion evolution:

$$n_{\text{T}}(x, y, t) = A_{\text{T}}(t)e^{-\left[\frac{x-x_0^{\text{T}}(t)}{\sigma_x^{\text{T}}(t)}\right]^2-\left[\frac{y-y_0^{\text{T}}(t)}{\sigma_y^{\text{T}}(t)}\right]^2} \tag{2.73}$$

where experimentally $x_0^{\text{T}}(t)$, $y_0^{\text{T}}(t)$, $\sigma_x^{\text{T}}(t)$, $\sigma_y^{\text{T}}(t)$, $A_{\text{T}}(t)$ are free parameters of the integrated thermal density profile are fitted for each cloud. For the condensate the integrated Thomas-Fermi profile of a cloud trapped in a harmonic trap is used:

$$n_{\text{C}}(x, y, z) = \max\left[0, \frac{\mu - \frac{1}{2}m\left(\omega_x^2x^2 + \omega_y^2y^2 + \omega_z^2z^2\right)}{g}\right], \tag{2.74}$$

with the assumption that during TOF the cloud evolves such that $n_{\text{C}}(x, y, z, t) = a(t) * n_{\text{C}}(b(t) * (x - c(t)) + d(t) * (y - e(t)) + f(t) * (z - g(t))$ holds true then integrated column density of the cloud at any point in time can be described by

$$n_{\text{C}}(x, y, t) = A_{\text{C}}(t)\left(1 - \min\left[1, \left(\frac{x - x_0^{\text{C}}(t)}{\sigma_x^{\text{C}}(t)}\right)^2 + \left(\frac{y - y_0^{\text{C}}(t)}{\sigma_y^{\text{C}}(t)}\right)^2\right]\right)^{3/2} \tag{2.75}$$

where $x_0^{\text{C}}(t)$, $y_0^{\text{C}}(t)$, $A_{\text{C}}(t)$, $\sigma_x^{\text{C}}(t)$, $\sigma_y^{\text{C}}(t)$ are free parameters. The full cloud profile should be given by

$$n_{\text{Cloud}}(x, y, t) = n_{\text{T}}(x, y, t) + n_{\text{C}}(x, y, t) + c_0, \tag{2.76}$$

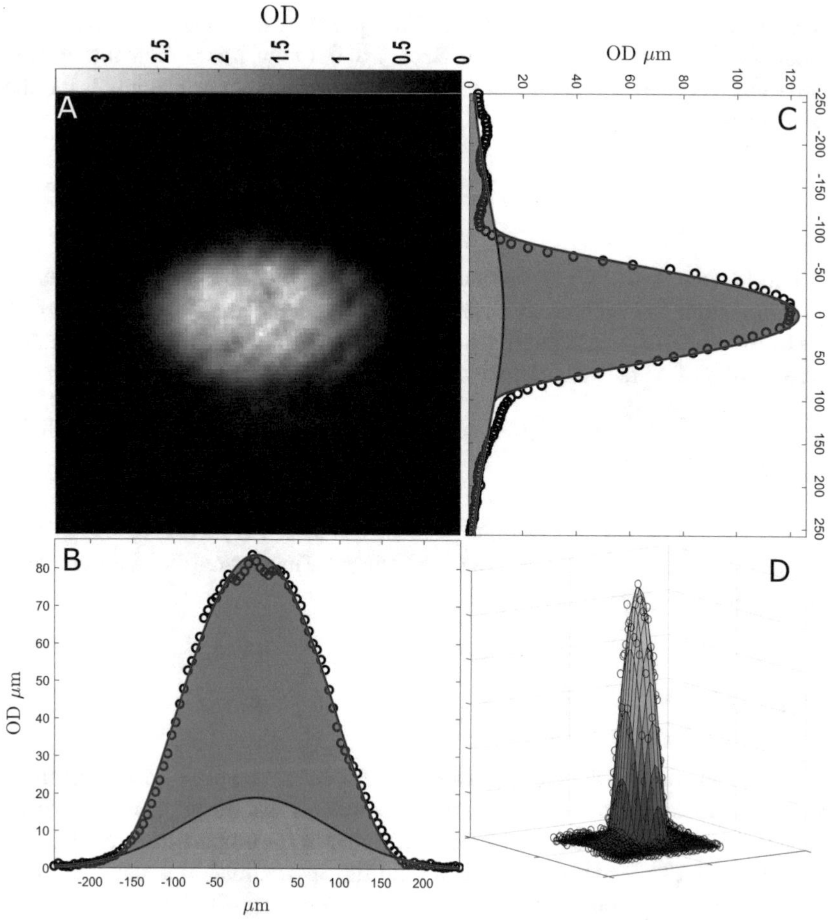

Fig. 2.4 Atom cloud fitting procedure. **A** shows the optical density of an atom cloud after 25 ms of free expansion. **B** shows the integrated optical density profile along the vertical axis with black representing the data point, shaded blue region representing the condensate atoms and shaded red region representing the thermal atoms. Both the thermal and condensate profiles are generated Eq. (2.76). **C** is the same as in **B** but for the horizontally integrated profile. **D** shows the fitted cloud density as a solid surface with the measure optical density depicted as the black dots

where c_0 is a constant offset included to account for the noise on the camera that possibly does not average to 0. In practise, $y_0^{\mathrm{T}} = y_0^{\mathrm{C}}$ and $x_0^{\mathrm{T}} = x_0^{\mathrm{C}}$ since for most systems the thermal and condensate clouds are not offset. An example fit is shown in Fig. 2.4.

The fitting used above in Eq. (2.76) works for an initial Gaussian cloud but can be quite inaccurate when trying to fit BECs which are initially confined in hard wall traps, as we discuss in later chapters, where a digital micromirror device (DMD) is used to provide trapping. To improve the accuracy of the fitting when doing TOF analysis of clouds which are trapped in DMD patterns, it is possible to model the trap potential $V(\vec{r})$, this is described in Sect. 4.6. For now, it is sufficient to assume the pre-release trapping potential is known. From this potential, the initial thermal cloud density is

$$n^0_{Thermal}(\vec{r}, N, T) = N \frac{\exp\left[-V(\vec{r})/(k_B T)\right]}{\int \exp\left[-V(\vec{r})/(k_B T)\right] d\vec{r}}. \tag{2.77}$$

for cloud temperature (T), and thermal atom number (N). The normalized thermal atom momentum ($\vec{p}$) probability distribution for a given energy cutoff (E_{Cutoff}) is given by the Boltzmann distribution normalized to have a unit area when integrating the probability of all the states above the energy cutoff

$$\mathrm{P}(\vec{p}, E_{\text{Cutoff}}, T) = \begin{cases} \left[2\sqrt{2}\pi m k_B T \left(2\exp\left[\frac{-E_{\text{Cutoff}}}{k_B T}\right] + \sqrt{\pi k_B m T}\,\mathrm{Erfc}\left[\sqrt{\frac{E_{\text{Cutoff}}}{k_B T}}\right]\right)\right]^{-1} \exp\left[\frac{-|\vec{p}|^2}{2mk_B T}\right] & \text{if} |\vec{p}| \geq \sqrt{2mE_{\text{Cutoff}}} \\ 0 & \text{otherwise.} \end{cases}$$

In the trap, atoms initially at position $\vec{r}$ will have at least $V(\vec{r})$ worth of energy once released from the trap which acts as a cutoff energy. The thermal cloud density after a time evolution is given by

$$n_{\text{Thermal}}(\vec{r}, t, N, T) = \int n^0_{\text{Thermal}}\left(\vec{r} - \frac{\vec{p}t}{m}, N, T\right) \mathrm{P}(\vec{p}, V(\vec{r}), T) d\vec{p}. \tag{2.78}$$

This integral is computationally extensive since it cannot be performed as a simple convolution. To decrease computation time, we assume that the momentum cutoff is nonexistent since the cutoff is (exponentially) inversely- proportional to the initial density, see Eq. (2.77). This allows us to find the approximate thermal cloud final density distribution by performing

$$n_{\text{Thermal}}(\vec{r}, t, N, T) \approx \int n^0_{\text{Thermal}}\left(\vec{r} - \frac{\vec{p}t}{m}, N, T\right) \mathrm{P}(\vec{p}, 0, T) d\vec{p}. \tag{2.79}$$

This convolution can be performed using fast Fourier transforms (FFTs).

To model the BEC, we start with the Thomas-Fermi wave function, given by Eq. (2.15), using an estimate of the chemical potential (μ). The simple method to find the wave function after TOF is to numerically solve the 3D GPE using the Runge-Kutta method. The computational resource needed to perform a parameter optimization on μ with this approach is impractical in terms of computational time during experimental runs. To reduce the computational time, we note that for BECs where the horizontal Thomas-Fermi widths ($x_{\text{TF}}, y_{\text{TF}}$) are substantially wider than the vertical Thomas-Fermi width (z_{TF}) most of the interaction energy will contribute to the expansion in the vertical direction. This is due to having a broader initial vertical momentum distribution compared to the horizontal distribution. Also, having a more uniform BEC in the horizontal direction, due to the trap being flatter horizontally, further increases the fraction of the interaction energy which goes to the vertical expansion. Since the DMD does not cover the entire trap (see Sect. 4.6), loading of the atoms can result in condensates inside and outside of the DMD pattern. The horizontal wave function is thus modelled as 2 BECs which do not overlap: one trapped inside the DMD and one trapped outside of the DMD pattern. This leads to a wave function which can be well approximated by:

$$\psi_{\text{BEC}}(\vec{r}, \mu, 0) \approx \frac{\left[\psi_{\text{TF}}^{\text{DMD}}(x, y, 0, \mu) + \psi_{\text{TF}}^{\text{OutDMD}}(x, y, 0, \mu)\right] \psi_{\text{TF}}^{z}(0, 0, z, \mu, 0)}{\psi_{\text{TF}}(0, 0, 0, \mu, 0)}, \tag{2.80}$$

where ψ_{TF}^{z} is the Thomas-Fermi wave functions for $V(0, 0, z)$ and $\psi_{\text{TF}}^{\text{DMD}}$ ($\psi_{\text{TF}}^{\text{OutDMD}}$) are the Thomas-Fermi wave functions for $V(x, y, 0)$ accounting for the atoms inside (outside) the DMD pattern. The wave function is normalized to ensure our reconstructed function approximates the actual Thomas-Fermi wave function in amplitude and dimensions. Because the trapping frequency is larger in the vertical than in the horizontal direction most of the interaction energy will go towards the vertical velocity profile. Therefore the horizontal and vertical wave function evolutions will be different, the latter has to be done by numerically solving the GPE [Eq. (2.3)] due to the interaction energy; whereas, the former can be done in three steps by going to momentum space

$$\Phi\left(k_x, k_y, \mu, 0\right) = \frac{1}{\sqrt{2\pi}} \int\int \left[\psi_{\text{TF}}^{\text{DMD}}(x, y, 0, \mu) + \psi_{\text{TF}}^{\text{OutDMD}}(x, y, 0, \mu)\right] e^{-i(k_x x + k_y y)} dx dy, \tag{2.81}$$

evolving the wave function

$$\Phi\left(k_x, k_y, \mu, t\right) = \Phi\left(k_x, k_y, \mu, 0\right) * \exp\left[-i\frac{\hbar(k_x^2 + k_y^2)t}{2m}\right], \tag{2.82}$$

and taking the inverse Fourier transform to obtain the time evolved horizontal wave function in real space

$$\psi_H\left(x, y, \mu, t\right) = \frac{1}{\sqrt{2\pi}} \int \Phi\left(k_x, k_y, \mu, t\right) e^{i(k_x x + k_y y)} dk_x dk_y. \tag{2.83}$$

For the vertical direction, we use Runge-Kutta method to evolve the wave function $\psi_{\text{TF}}^{z}(0, 0, z, \mu, 0)$ under the GPE which results in $\psi_z(z, \mu, t)$. The TOF evolved BEC wave function is then given by:

$$\psi_{BEC}(\vec{r}, \mu, t) \approx \psi_H(x, y, \mu, t)\psi_z(0, 0, z, \mu, t), \tag{2.84}$$

and the density is given by

$$n_{\mathrm{BEC}}(\vec{r}, \mu, t) = |\psi_{BEC}(\vec{r}, \mu, t)|^2 . \tag{2.85}$$

The camera images the integrated density along the y-direction which means that the signal we expect to measure is given by

$$n_{Total}(x, y, \mu, t, N, T, c0, \vec{r}_c) = c_0 + \int \left[n_{\mathrm{BEC}}(\vec{r} - \vec{r}_c, \mu, t) + n_{\mathrm{Thermal}}(\vec{r} - \vec{r}_c, t, N, T)\right] dy \tag{2.86}$$

where c_0 is a constant offset included to account for the noise on the camera that possibly does not average to 0, and $\vec{r}_c$ is the centre offset position with 2 degrees of freedom since the y-direction is integrated out. This leaves us with a fit with 6 degrees of freedom which are the chemical potential (μ), the thermal atom number (N), the thermal cloud temperature (T), the noise offset (c_0), the cloud centre ($\vec{r}_c = x_c\hat{x} + z_c\hat{z}$). The time evolution is a known quantity which is set experimentally and is therefore not a free parameter. An example fit is shown in Fig. 2.5, where a cloud initially trapped in a 50 μm radius circle is expanded in TOF for 100 ms and then fit using Eq. (2.86). The advantages of using Eq. (2.86) as opposed to Eq. (2.76) are that it is a more accurate model, it requires fewer fit parameters and it directly fits physically relevant parameters which need to be extracted when using Eq. (2.76). This comes at the small cost of having to have an accurate model of the trapping potential and increased computational time.

2.5.3 Faraday Imaging

Another way to probe the cloud density is to use the phase information that is imprinted onto the probe beam after interacting with the cloud. This effect enables phase contrast imaging [47, 50] and Faraday imaging [51, 52]. One of the reasons to use Faraday imaging over absorption imaging is that the measured signal changes as the square of the column density (i.e. $I_{\mathrm{M}}^{\mathrm{Faraday}}(x, y)/I_{\mathrm{P}}^{\mathrm{Faraday}}(x, y) \propto n_{\mathrm{2D}}^2(x, y)$), making the signal sensitive to small fluctuations in the density, ideal for the detection of vortices. It also requires no optical pumping step which means that there is no blurring of the image due to radiation pressure during the repump and probe steps in absorption imaging, and the atoms are not pushed out of the plane of focus [53]. One of the drawbacks is a low signal when trying to measure low-density clouds, and the impracticality of measuring the probe beam light profile for every measurement which complicates extracting the actual atom numbers. Faraday imaging is the main imaging method used for vertically imaging the cloud in situ, but for optically thin clouds absorption imaging is the preferred technique.

When light passes through an atom cloud, the electric field interacts with the cloud, which acts as a lens with refractive index dependent on the local density [47]. Assuming that the cloud is small, in the probe beam propagation direction ($\hat{z}$), the phase shift acquired by the electric field $\vec{E}_0(x, y)$ is given by:

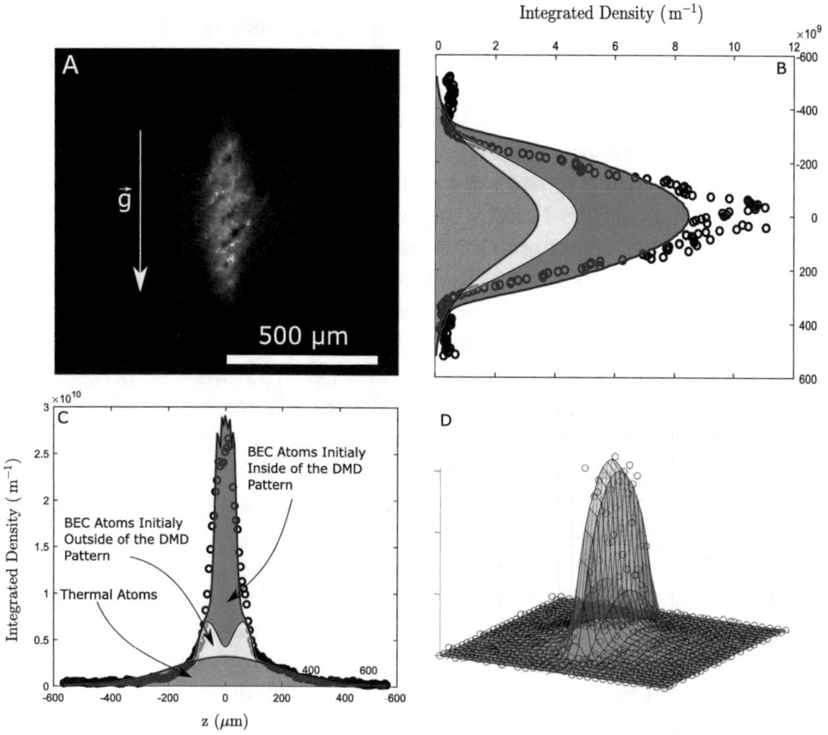

Fig. 2.5 Atom cloud fitting procedure for DMD trapped clouds. **A** shows the measured density of an atom cloud after 100 ms of free expansion with the arrow representing the direction of gravity. **B** shows the integrated optical density profile along the horizontal axis with black representing the data point, shaded blue (yellow) region representing the condensate atoms initially inside (outside) the DMD pattern and shaded red region representing the thermal atoms. Both the thermal and condensate profiles are generated Eq. (2.86). **C** is the same as in **B** but for the vertical direction. **D** shows the fitted cloud density as a solid surface with the measured density depicted as the black dots

$$\vec{E}(\vec{r}) = \vec{E}_0(x, y)e^{i\alpha(\vec{r})} = \vec{E}_0(x, y)\exp\left[-i\sigma_0 \int_0^z n(\vec{r})dz \frac{\Delta/\Gamma}{1 + 4(\Delta/\Gamma)^2 + \frac{I(\vec{r})}{I_{sat}}}\right] \tag{2.87}$$

where I_{sat} is the saturation intensity given by Eq. (2.68), σ_0 is the cross section given by Eq. (2.67), Γ is the transition linewidth, $\Delta = \omega - \omega_0$ is the detuning of the probe beam angular frequency (ω) from the transition frequency ω_0 and $n(\vec{r})$ is the 3D density. After passing through the atomic cloud the probe beam electric field's phase evolves by

$$\alpha(x, y) = -\sigma_0 n_{2D}(x, y) \frac{\Delta/\Gamma}{1 + 4(\Delta/\Gamma)^2 + I(\vec{r})/I_{sat}}. \tag{2.88}$$

The key to understanding Faraday imaging is that linearly polarized ($\hat{\pi}$) light can be decomposed into a superposition of circularly polarized light components, $\hat{\pi} = (\sigma_+ - \sigma_-)/\sqrt{2}$ [52]. Assuming that the atomic dipole moment $\vec{d}$ is well defined relative to the polarization vector $\hat{\varepsilon}$ of the probe beam, due to an externally applied magnetic field, it is possible for each of the circulations to experience a different phase shift when passing through the cloud. This leads to a rotation of the linearly polarized light by the Faraday angle $\theta_{\text{F}}(x, y)$. Using a polarizer, the Faraday rotation can be mapped onto an intensity fluctuation, through Malus's law, $I_{\text{M}}(x, y) = I_{\text{P}}(x, y)\cos^2(\theta_{\text{F}}(x, y) + \theta_0)$ where θ_0 is the initial angle of the light to the polarizer's polarization axis. The intensity fluctuations can then be measured by a camera. The Faraday angle is simply the difference of the imprinted phase on the circular polarizations, $\theta_{\text{F}}(x, y) = \phi_+(x, y) - \phi_-(x, y)$ where $\phi_\pm(x, y)$ are the phases imprinted onto the $\sigma_\pm$ components of the probe beam respectively.

Faraday images are taken using the set-up shown in Fig. 2.6 where a beam of P-polarised light is sent through the atom cloud. A half-wave plate is used to rotate the polarization by $\pi/2$ to S-polarisation and then a polarising beam-splitting cube is used as a polariser to map the Faraday rotation angle to an intensity. The camera is positioned to image the 'rotated light' that is transmitted, so there is minimal signal without the presence of the atoms Fig. 2.6A. With the atoms present, the light undergoes a Faraday angle shift. Since the original polarisation was perpendicular relative to the polarizer axis, $\theta_0 = \pi/2$ and $I_{\text{M}}(x, y) = I_{\text{P}}(x, y)\sin^2(\theta_{\text{F}}(x, y)$ results. The light signal will be dependent on the local atom column density the probe light had to travel through, as shown in the inset of Fig. 2.6B. The half-wave plate can also be rotated to do absorption imaging as in Fig. 2.1A. Wave plate angles inbetween can also be used to get phase-contrast imaging, where part of the transmitted probe beam interacts with the rotated component that has coupled to the $\sigma_\pm$ optical transitions acquiring a phase shift [50].

Figure 2.7A shows the allowed optical transitions for a cloud initially in the $5^2\text{S}_{1/2}|\text{F} = 1, m_{\text{F}} = -1\rangle$ using a $\hat{\pi}$ laser detuned from the $5^2\text{S}_{1/2}|\text{F} = 1\rangle \rightarrow 5^2\text{P}_{3/2}|\text{F}' = 2\rangle$ transition by Δ, in the absence of a magnetic field, showing that even in the presence of a very small magnetic field the cloud is still birefringent. Faraday images are usually performed in the levitation field (Sect. 3.3.1), which has an 80 G field component along the imaging axis direction, leading to a Zeeman energy shift $\Delta E = \mu_{\text{B}} m_{\text{F}} g_{\text{F}} B_z$ of the sub-levels of the hyperfine states.

Using Eqs. (2.67), (2.68) and (2.88) one can calculate the total phase imprinted onto the different circulations by summing over the phase accumulated through interacting with the atoms, for all possible atomic transitions

$$\phi_\pm = -n_{2\text{D}}(x, y) \sum_{\text{F}'=\max(\text{F}-1,|m_{\text{F}}\pm 1|)}^{\text{F}+1} \sigma_0^{\text{F}'} \frac{\Delta_{\text{F}'}/\Gamma}{1 + 4(\Delta_{\text{F}'}/\Gamma)^2 + I(\vec{r})/I_{\text{sat}}^{\text{F}'}}, \tag{2.89}$$

where F and m_{F} are the hyperfine levels and sub-levels of the atomic cloud to be probed. $I_{\text{sat}}^{\text{F}'}$, $\sigma_0^{\text{F}'}$ are given by Eqs. (2.68) and (2.67) for the respective transition,

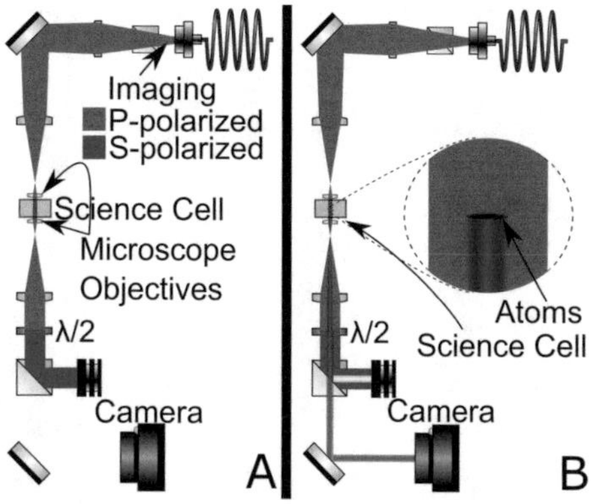

Fig. 2.6 Faraday imaging light **A** without the atoms, where all of the probe light is dumped by the probing polarizing beam splitter, in contrast with Fig. 2.1A where the waveplate is arranged to let all the light through to allow for absorption imaging. In the presence of atoms **B** the birefringence of the cloud rotates part of the probe light, and the camera images the rotated light which can be mapped back to an atom density using Malus's law and Eq. (2.89)

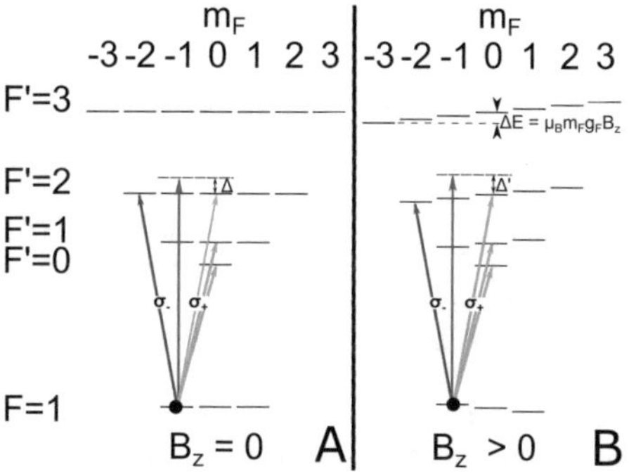

Fig. 2.7 Optical process behind Faraday imaging. Hyperfine levels and their internal sub-levels for ^{87}Rb on the D_2 line without the presence of a magnetic field (**A**) and with a magnetic field (**B**) assuming only linear Zeeman shifts of the internal sub-levels. The atoms are initially in the $5^2S_{1/2}|F = 1, m_F = -1\rangle$ state represented by the black dot. The probe laser (red arrow) is detuned from the $5^2S_{1/2}|F = 1\rangle \rightarrow 5^2P_{3/2}|F' = 2\rangle$ transition by (Δ). The orange (blue) arrows represent the allowed transitions for an atom absorbing $\sigma_+(\sigma_-)$ light which makes up the linearly polarized light. Note that only optical $F' = F\pm1$ is allowed due to angular momentum conservation

$5^2S_{1/2}|F\rangle \rightarrow 5^2P_{3/2}|F'\rangle$ with polarization vector $\hat{\sigma}_\pm$ and $\Delta_{F'}$ is the detuning of the laser from the $5^2S_{1/2}|F\rangle \rightarrow 5^2P_{3/2}|F'\rangle$ accounting for the relative Zeeman shift of the initial and final sub-levels given by $\Delta_Z = \mu_B(m_{F'}g_{F'} - m_F g_F)B_z/\hbar$. This formula assumes that all the atoms stay in their initial state throughout the interaction, and that the local densities are unaffected by the probe beam. Figure 2.8 shows the expected Faraday angle and signal for a low magnetic field case ($B_z = 0.1$ G) and the magnetic field present at the atom location during levitation ($B_z = 80$ G). Experimentally, a detuning of $\Delta/2\pi = -196$ MHz produces the best signal in the levitation field.

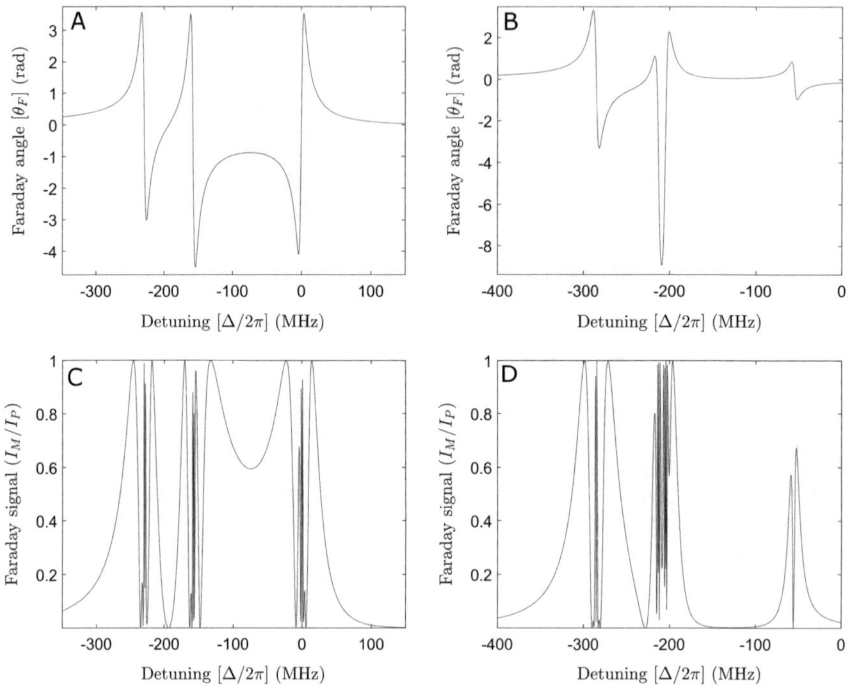

Fig. 2.8 Plots of the typical expected Faraday rotation (θ_F) and signal (I_M/I_P) for a uniform atom column density $n_{2D}(\vec{r}) = 3.1 \times 10^{14}$ m^{-2}, and probe beam uniform intensity $I_P(\vec{r})$ = 15 W/m^2. Generated using the method presented in [52]. **A** and **B** show the expected Faraday angular rotation of the probe electric field when passing through the cloud for a magnetic field aligned with the imaging beam of $B_z = 0.1$ G and our levitation magnetic field at the atoms of $B_z = 80$ G, respectively. **C** and **D** show the expected signal for the same condition as **A** and **B**, respectively with the assumption that the initial incident light is perpendicular to the polarizer polarization axis. The for both cases the detuning is measured from the $5^2S_{1/2}|F = 1\rangle \rightarrow 5^2P_{3/2}|F' = 2\rangle$ (D_2) transition as shown in Fig. 2.7A

The Faraday angle can be related to the density through using $\theta_F(x, y) = n_{2D}(x, y)(\sigma^+_\Delta - \sigma^-_\Delta) = n_{2D}(x, y)\sigma_{eff}$, where $\sigma^\pm$ is obtained by dividing Eq. (2.89) by $n_{2D}(x, y)$ and σ^Δ_{eff} is the effective angular rotation cross section for a given detuning Δ. Using Malus's law with $\theta_0 = \pi/2$ allows for the expression of the density in terms of the probe beam intensity profile $I_P(x, y)$ and the measured intensity profile $I_M(x, y)$ as

$$n_{2D}(x, y) = \frac{1}{\sigma^\Delta_{eff}} \arcsin\left[\sqrt{\frac{I_M(x, y)}{I_P(x, y)}}\right]. \tag{2.90}$$

The probe beam intensity profile cannot be measured during the experiment since our wave-plate is stationary. Therefore, $I_P(x, y)$ needs to be measured either before or after the experimental run. Two images are taken during the experiment one of the probe light without $I_{wo}(x, y)$ and one with $I_w(x, y)$ the atoms Fig. 2.6A, B, respectively. The measured intensity profile is given by $I_M(x, y) = I_w(x, y) - I_{wo}(x, y)$.

There are two main regimes for the noise in Faraday images. When the signal is large compared to the noise, $I_M(x, y) \gg p_G(x, y)$ where $p_G(x, y)$ is the rms noise local noise, the noise on the atom count will tend to average to zero when integrating over large areas. However, if the rms noise is comparable or bigger that the signal (true for regions where there are no atoms) then a problem arises in that $I_M(x, y)$ can be negative, leading to imaginary roots in Eq. (2.90). To remedy this, it can be noted that in the limit where there are no atoms the noise will be symmetric such that half the values will be negative and half will be positive with so instead of ignoring the imaginary root as non-physical they can be treated as negative real atom counts so that the noise averages out over larger areas, leading to:

$$n_{2D}(x, y) = \frac{1}{\sigma_\Delta^{\text{eff}}} \arcsin\left[\text{Re}\left[\sqrt{\frac{I_M(x, y)}{I_P(x, y)}}\right] - \text{Im}\left[\sqrt{\frac{I_M(x, y)}{I_P(x, y)}}\right]\right], \tag{2.91}$$

which partially compensates for the experimental noise.

2.5.4 *Momentum Spectroscopy Through Bragg-Scattering*

Bragg spectroscopy has been used to make a variety of measurements in cold atoms. It has been used to measure the coherence length of a Bose-Einstein condensate [54], the momentum distribution of cold atom gases [55], the superfluid pairing gap and speed of sound in strongly interacting Fermi gases [56], the circulation sign of different vortices [57], and much more. In particular, in this thesis Bragg-scattering is used to measure the 1D velocity distribution of the BEC, allowing us to infer the sign of the vortices (Sect. 6.5).

To Bragg-scatter an atom cloud, two phase-locked lasers with wavevectors $\vec{k}_1$, and $\vec{k}_2$ and frequencies ω_1, and ω_2 are made to intercept at the cloud location, as shown in Fig. 2.9. A single atom scatters by spontaneously absorbing a photon from on of the beams and then remitting a stimulated photon into the other beam changing its momentum and energy in the process. This process can be repeated N times, known as high-order Bragg scattering.

The scattering process of the photons with the atom can be seen as a collision, which requires conservation of energy and momentum which can be satisfied through tuning the frequency difference of the two Bragg lasers. Conservation of energy requires the energy change of the atom during the process be equal to the energy difference between the absorbed and re-emitted photons:

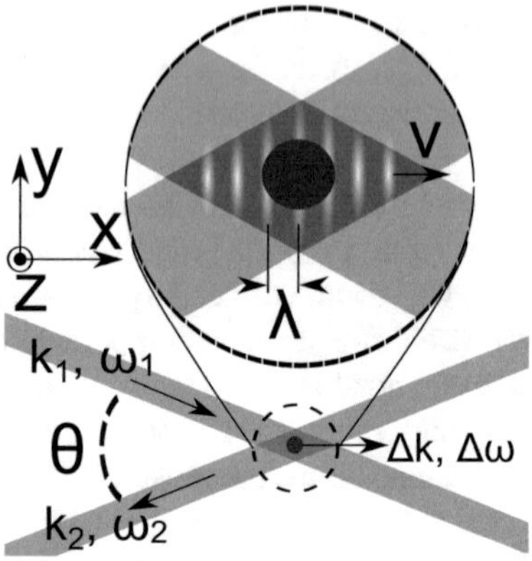

Fig. 2.9 Depiction of a Bragg diffraction process where two phase-locked laser beams (red) of linear polarization in $\hat{z}$-direction, wavevector ($\vec{k}_1$,$\vec{k}_2$) and frequency (ω_1, ω_2) are made to intercept with angles ($-\theta/2$,$\pi - \theta/2$) relative to the x-axis. A lattice, of wavelength λ, moving at v in the $\hat{x}$-direction, assuming that $\omega_1 > \omega_2$, is created. If the interception of these two beams is made to occur at the cloud location (blue), and an atom in the cloud undergoes Bragg-scattering, its change in energy is given by $\Delta E = \hbar\Delta\omega$ and change in momentum by ($\Delta\vec{P} = \hbar\Delta\vec{k}$)

$$\frac{\hbar^2}{2m}\left[|\vec{k}_{\text{out}}^{\text{atom}}|^2 - |\vec{k}_{\text{in}}^{\text{atom}}|^2\right] = N\hbar\left(\omega_2 - \omega_1\right). \tag{2.92}$$

The momentum conservation simply means that wavenumbers must be conserved

$$\vec{k}_{\text{in}}^{\text{atom}} - \vec{k}_{\text{out}}^{\text{atom}} = N\Delta\vec{k}^{\text{photon}} = 2N|\vec{k}^{\text{photon}}|\cos(\theta/2)\hat{x}, \tag{2.93}$$

where we have assumed that $|\vec{k}_{\text{in}}^{\text{photon}}| = |\vec{k}_{\text{out}}^{\text{photon}}| = |\vec{k}^{\text{photon}}|$, since the detuning is small relative to the frequency of the beam. Combining Eqs. (2.92) and (2.93) we get the Bragg beam frequency detuning resonance condition which is

$$\delta = \delta_0 + \delta_v = \frac{N\hbar|\vec{k}^{\text{photon}}|^2\cos^2(\theta/2)}{\pi m} + \frac{\hbar|\vec{k}^{\text{photon}}|\cos(\theta/2)\vec{k}^{\text{atom}}\cdot\hat{x}}{\pi m}. \tag{2.94}$$

$\delta_0 = 2\pi N \times 15.08$ kHz is a fixed detuning that conserves energy and momentum during the scattering process for atom initially at rest, while δ_v is a detuning due to the initial velocity of the atom in the direction of the scattering ($\hat{x}$). The recoil velocity is the velocity imparted to the atom by the scattering process and gives:

$$\vec{v}_{\text{recoil}}^{\text{atom}} = \vec{v}_{\text{out}}^{\text{atom}} - \vec{v}_{\text{in}}^{\text{atom}} = \frac{2Nhf}{mc}\cos(\theta/2)\hat{x} \tag{2.95}$$

where f is the frequency of the Bragg laser.

The specific set-up in terms of angles and laser frequencies utilised in our experiments can be found in Sect. 3.4.1.

References

1. Guénault T (1995) Statistical physics. Springer, Netherlands
2. Bose (1924) Plancks Gesetz und Lichtquantenhypothese. Zeitschrift für Physik 26:178–181
3. Einstein A (1924) Quantentheorie des einatomigen idealen Gases. Sitzungsberichte der Preussischen Akadamie der Wissenschaften, Physikalisch-mathematische Klasse 1
4. Einstein A (1925) Quantentheorie des einatomigen idealen Gases: Zweite Abhandlung. Sitzberichte der Preussischen Akadamie der Wissenschaften, Physikalisch-mathematische Klasse 1
5. Pitaevsii LP (1961) Vortex lines in an imperfect bose gas. Sov Phys J Exp Theor Phys 40:646–651
6. Gross EP (1961) Structure of a quantized vortex in boson systems. Il Nuovo Cimento 1955–1965(20):454–477
7. Gross EP (1963) Hydrodynamics of a superfluid condensate. J Math Phys 4:195–207
8. Barenghi CF, L'vov VS, Roche P-E (2014) Experimental, numerical, and analytical velocity spectra in turbulent quantum fluid. Proc Natl Acad Sci 111:4683–4690
9. Pethick CJ, Smith H (2008) Bose-Einstein condensation in dilute gases, 2nd edn. Cambridge University Press, Cambridge
10. Davis MJ (2001) Dynamics of Bose-Einsein condensation. University of Oxford
11. Feshbach H (1958) Unified theory of nuclear reactions. Ann Phys 5:357–390
12. Feshbach H (1962) A unified theory of nuclear reactions. II. Ann Phys 19:287–313
13. Chin C, Grimm R, Julienne P, Tiesinga E (2010) Feshbach resonances in ultracold gases. Rev Mod Phys 82:1225–1286
14. Bao W, Du Q (2004) Computing the ground state solution of Bose-Einstein condensates by a normalized gradient flow. SIAM J Sci Comput 25:1674–1697
15. Bell TA et al (2018) Phase and micromotion of Bose-Einstein condensates in a time-averaged ring trap. Phys Rev A 98:013604
16. Reeves M (2017) Quantum analogues of two-dimensional classical turbulence. University of Otago
17. Bogolubov N (1947) On the theory of superfluidity. J Phys XI:23–32
18. Lee TD, Huang K, Yang CN (1957) Eigenvalues and eigenfunctions of a bose system of hard spheres and its low-temperature properties. Phys Rev 106:1135–1145
19. Landau L (1941) Theory of the superfluidity of helium II. Phys Rev 60:356–358
20. Groszek AJ (2018) Vortex dynamics in two-dimensional Bose-Einstein condensates doctor of philosophy. Monash University
21. Lundh E (2002) Multiply quantized vortices in trapped Bose-Einstein condensates. Phys Rev A 65:043604
22. Huhtamäki JAM, Möttönen M, Virtanen SMM (2006) Dynamically stable multiply quantized vortices in dilute Bose-Einstein condensates. Phys Rev A 74:063619
23. Rooney SJ, Blakie PB, Anderson BP, Bradley AS (2011) Suppression of Kelvon-induced decay of quantized vortices in oblate Bose-Einstein condensates. Phys. Rev. A 84:023637
24. Fetter AL (1966) Vortices in an imperfect bose gas. IV. Translational velocity. Phys Rev 151:100–104
25. Creswick RJ, Morrison HL (1980) On the dynamics of quantum vortices. Phys Lett A 76:267–268
26. Neu JC (1990) Vortices in complex scalar fields. Phys D: Nonlinear Phenom 43:385–406
27. Lund F (1991) Defect dynamics for the nonlinear Schrödinger equation derived from a variational principle. Phys Lett A 159:245–251
28. Kawasaki K (1984) Defect-phase dynamics for dissipative media with potential. Prog Theor Phys Suppl 80:123–138
29. Lucas A, Surowka P (2014) Sound-induced vortex interactions in a zero-temperature two-dimensional superfluid. Phys Rev A 90:053617
30. Bustamante MD, Nazarenko S (2015) Derivation of the Biot-Savart equation from the nonlinear Schrödinger equation. Phys Rev E 92:053019

31. Törnkvist O, Schröder E (1997) Vortex dynamics in dissipative systems. Phys Rev Lett 78:1908
32. Nore C, Abid M, Brachet ME (1997) Kolmogorov turbulence in low-temperature superflows. Phys Rev Lett 78:3896–3899
33. Aref H (1983) Integrable, chaotic, and turbulent vortex motion in two-dimensional flows. Ann Rev Fluid Mech 15:345–389
34. Newton PK (2013)) The N-vortex problem: analytical techniques. Springer Science & Business Media
35. Lin CC (1941) On the motion of vortices in two dimensions. Proc Natl Acad Scie U S A 27:570–575
36. Griffiths DJ (2012) Introduction to electrodynamics, 4th edn. Cambridge University Press, Cambridge
37. Kraichnan RH, Montgomery D (1980) Two-dimensional turbulence. Rep Prog Phys 43:547
38. Pointin YB, Lundgren TS (1976) Statistical mechanics of two-dimensional vortices in a bounded container. Phys Fluids 19:1459–1470
39. Eyink GL, Sreenivasan KR (2006) Onsager and the theory of hydrodynamic turbulence. Rev Mod Phys 78:87–135
40. Viecelli JA (1995) Equilibrium properties of the condensed states of a turbulent two-dimensional neutral vortex system. Phys Fluids 7:1402–1417
41. Kober, H (1957) Dictionary of conformal representations. Dover New York
42. Grimm R, Weidemüller M, Ovchinnikov YB (2000) In: Bederson B, Walther H (eds) Academic Press, pp 95–170
43. Denschlag J et al (2000) Generating solitons by phase engineering of a Bose-Einstein condensate. Science 287:97–101
44. Matthews MR et al (1998) Dynamical response of a Bose-Einstein condensate to a discontinuous change in internal state. Phys Rev Lett 81:243–247
45. Freilich DV, Bianchi DM, Kaufman AM, Langin TK, Hall DS (2010) Real-time dynamics of single vortex lines and vortex dipoles in a Bose-Einstein condensate. Science 329:1182–1185
46. Steck DA (2015) Rubidium 87 D Line Data 2.1.5
47. Ketterle W, Durfee D, Stamper-Kurn DM (1999) Bose-Einstein condensation in atomic gases 67. In: Proceedings of international school of physics "enrico fermi", vol CXL. IOS Press, Amsterdam
48. Metcalf HJ, Straten Pvd (2003) Laser cooling and trapping of atoms. J Opt Soc Am B 20:887–908
49. Reinaudi G, Lahaye T, Wang Z, Guéry-Odelin D (2007) Strong saturation absorption imaging of dense clouds of ultracold atoms. Opt Lett 32:3143–3145
50. Bradley CC, Sackett CA, Hulet RG (1997) Bose-Einstein condensation of lithium: observation of limited condensate number. Phys Rev Lett 78:985–989
51. Gajdacz M et al (2013) Non-destructive Faraday imaging of dynamically controlled ultracold atoms. Rev Sci Instrum 84:083105
52. Wilson KE (2015) Developing a toolkit for experimental studies of two-dimensional quantum turbulence in Bose-Einstein condensates. The University of Arizona
53. Hume DB et al (2013) Accurate atom counting in mesoscopic ensembles. Phys Rev Lett 111:253001
54. Stenger J et al (1999) Bragg spectroscopy of a Bose-Einstein condensate. Phys Rev Lett 82:4569–4573
55. Navon N, Gaunt AL, Smith RP, Hadzibabic Z (2016) Emergence of a turbulent cascade in a quantum gas. Nature 539:72–75
56. Hoinka S et al (2017) Goldstone mode and pair-breaking excitations in atomic Fermi superfluids. Nat Phys 13:943
57. Seo SW, Ko B, Kim JH, Shin Y (2017) Observation of vortex-antivortex pairing in decaying 2D turbulence of a superfluid gas. Sci Rep 7:4587

Chapter 3
A Versatile BEC Apparatus

3.1 Introduction

In this chapter, we give an overview of the experimental components of the University of Queensland BEC apparatus that have been modified and added to the experiment during my PhD. A typical setup for creation of Bose-Einstein condensate (BEC) consists of a vacuum system, magnetic coils with driving electronic circuits, and lasers for cooling and trapping. A well functioning ^{87}Rb BEC experiment will produce BEC of $\sim 0.5 - 5 \times 10^6$ atoms with lifetimes >10 s. We focus on experimental details that are relevant to the experiments presented in this thesis. These include the vacuum, electrical and laser systems which were all modified during my PhD. The evaporation procedure used to create a Bose-Einstein condensate from thermal atoms is summarised. Readers may refer to Nicholas McKay Parry's thesis [1] for a more complete description of the original design and construction of the experimental apparatus. In his thesis Nick describes the construction of a *Cicero*-controlled Opal Kelly FPGA driven NI PXI system, the distribution board for the lasers used for imaging, trapping, cooling in the experiment. Nick also constructed the 2D and 3D magneto-optical trap (MOT) and the transfer coils used to transfer the atoms from the 3DMOT to the science cell. The culmination of his work led to a usable cooling sequence which was used to achieve the first BEC in our lab.

3.2 Vacuum System

Our apparatus consists of a three-part vacuum system, comprising a two-dimensional magneto-optical trap (2DMOT), three-dimensional MOT (3DMOT) and attached Science Cell cuvette (see Fig. 3.1). The Science Cell is made of Suprasil quartz and has 1.25 mm thick walls with an external broadband anti-reflection coating. The 2DMOT is separated from the 3DMOT and science cell sections with a 12 mm long

© The Editor(s) (if applicable) and The Author(s), under exclusive license to Springer Nature Switzerland AG 2020

G. Guillaume, *Transport and Turbulence in Quasi-Uniform and Versatile Bose-Einstein Condensates*, Springer Theses,
https://doi.org/10.1007/978-3-030-54967-1_3

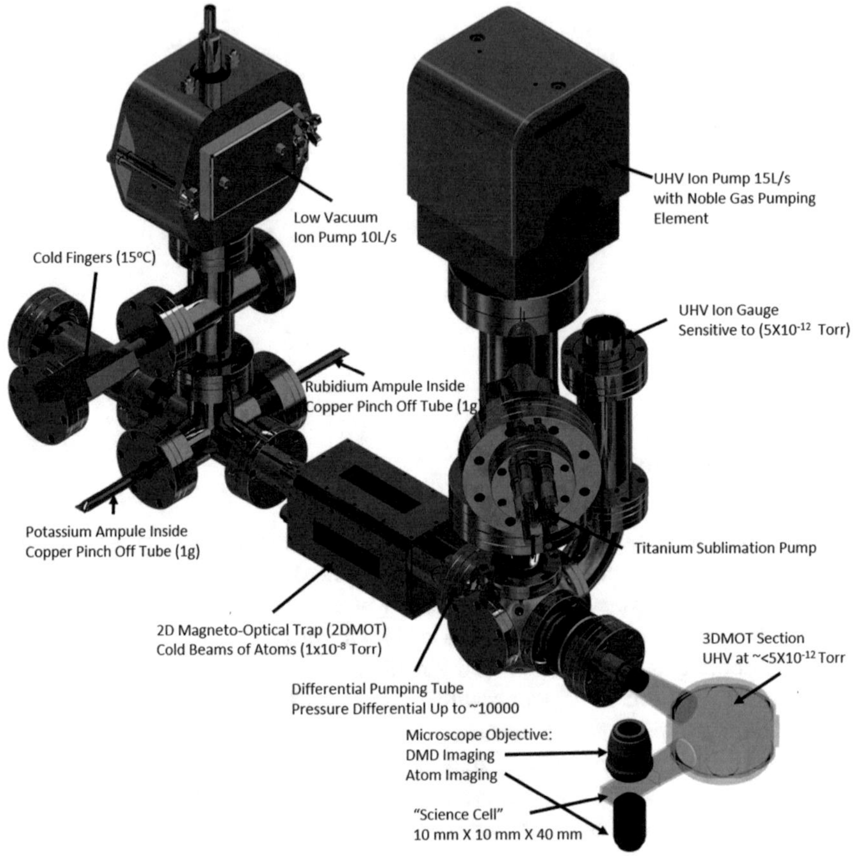

Fig. 3.1 Picture of the vacuum system used to run the experiments. The system consists of two main parts a low vacuum 2DMOT side with a pressure of around 10^{-8} Torr and a high vacuum 3DMOT/science cell side with a pressure of $<5 \times 10^{-12}$ Torr and they are separated by a differential pumping tube which is simply a small pinhole with low conductance see [1] for dimensions and specifications. The low vacuum side consists of an ion pump, two heated pinched of copper tubes with potassium and rubidium ampules inside them and a set of cold fingers cooled to 15°C. On the high vacuum side, we have a home made Ti-sub pump, combined with an Ion pump to pump the system, to measure the pressure we also have a UHV ion gauge. These are connected to a glass octagon inside of which a 3DMOT is generated, with a glass cell protrusion for our high-resolution optical access needs

differential pumping tube with a 1.2 mm inner diameter. This results in a pressure differential of ~15,000, with a pressure of $\sim 5 \times 10^{-12}$ Torr and 10^{-8} Torr on the 3DMOT and 2DMOT sides, respectively.

The low vacuum (10^{-8} Torr) side consists of a pair of cold fingers kept at 15°C which acts as a sink for the rubidium atoms in the system (it also includes a 10 L/s ion vacuum pump which we are not running as it succumbed to poisoning from the rubidium). There are two ampules with 1 g solid mass of Rb and K sitting inside a pinched off copper tubes on either side of the experiment. The K ampule was included for future expansion to a dual BEC of ^{87}Rb and ^{41}K. We warm up the Rb mass to a constant temperature to keep the Rb pressure constant in the system to create a constant supply of ^{87}Rb atoms. A 2DMOT square cell with four windows allows us to do two-axis cooling and a window on the end of the experiment allows for optical access for a push beam to push the atoms through the differential tube into the high vacuum side and to the science cell.

The ultra-high vacuum (UHV) is actively pumped by a 15 L/s ion pump with noble gas pumping element and a home-made titanium sublimation pump which is run once every 6–8 months (a UHV ion gauge was also added to the system but it is not used due to insufficient out-gassing during the baking process of the vacuum). The UHV measured pressure is thus degraded from out-gassing of the gauge when turned on, increasing the pressure that the gauge is trying to measure. The glass octagon gives good optical access for 3DMOT cooling and the science glass cell allows for high-resolution imaging and projection.

The initial experimental work in my PhD involved rebuilding the vacuum system to its current form. This was done because of the presence of a leak at the differential pumping tube on the low vacuum side of the system. This leak caused high pressure on both side of the system which meant low lifetimes and high thermal fractions in our small condensates preventing any useful experimentation. Details of the system bakeout were similar to that of Ref. [1] which can be summed up as a two weeks out-gassing of the system at high temperature to improve the vacuum to its current condition.

3.3 Electrical System

To create the magnetic fields needed for the 3DMOT, microwave (MW)-evaporation, and to transfer the atoms from the 3DMOT to the Science Cell, the system is equipped with three main quadrupole coil pairs depicted in Fig. 3.2. The MOT coils are used to create the needed magnetic field during the loading of the 3DMOT and the magnetic re-trapping (see Sects. 3.5.1 and 3.5.3). The transfer coil pair is only used in conjecture with the other two coils to transfer the atoms from the octagon glass cell to the science cell.

The BEC coil pair is the only coil pair that can be driven as either a Helmholtz, anti-Helmholtz pair or independently of one another due to the coil driving circuit as shown in Fig. 3.3. We achieve this by using a simple switching circuit to switch the polarity of the top coil and a by-pass current circuit to be able to take current away from the top coil allowing for an independent drive. This allows the BEC coils to create the gradient needed to perform MW-evaporation in our trap, but can also be used for accessing Feshbach resonances and to produce a levitation field.

One of the complications during the transfer from the magnetic field trap after MW-evaporation to the optical dipole trap was the drift of the zero position of the quadrupole magnetic field due to the background magnetic field coming from the ion

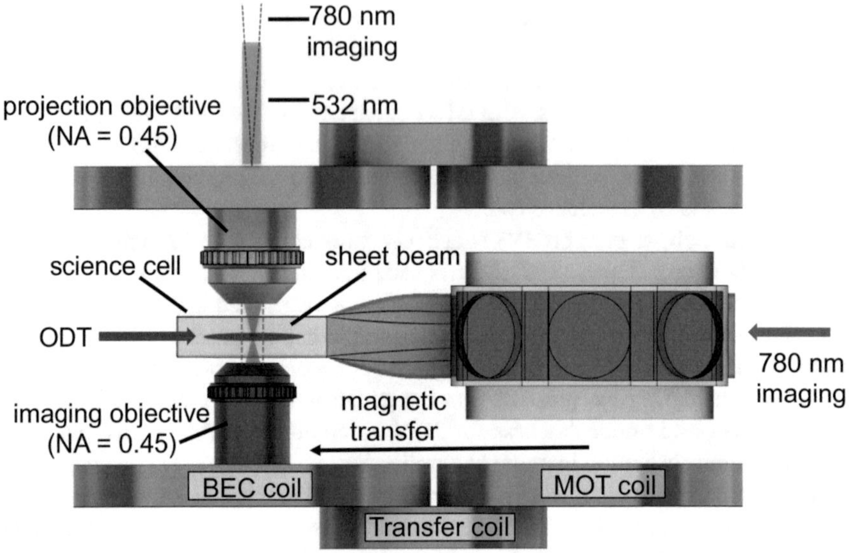

Fig. 3.2 Depiction of the main magnetic coils in the system. The MOT coils are used to create the 3DMOT and for magnetic trapping. The three coils field are then ramped in such a way as to transfer the atoms from the 3DMOT to the Science Cell, and then trapped inside the BEC coils magnetic field. A more thorough description of the system coil design can be found in [2]. The BEC coils are used for both trapping of the BEC in an anti-Helmholtz configuration, levitation of the cloud using a partial by-pass of the top coil and Feshbach resonances access in their Helmholtz configuration using the circuit shown in Fig. 3.3. The optical dipole trap, sheet beam, imaging lights and projected DMD light are shown for reference

pumps and earth's magnetic field. To cancel these fields we use a 3D printed zeroing coil assembly that fits around the science cell with 3 coil pairs perpendicular to one another to allow us to cancel out the magnetic field at the BEC location in all three directions. The vertical coils are driven as a Helmholtz pair and the other four coils are driven independently to allow for the cancellation of gradients and to allow us to bias the system. The 3D printed coil assembly is shown in Fig. 3.4 with the Science Cell and the microscope objectives also depicted. The side, BEC/MOT, vertical zeroing coils have 28, 54, 31 windings and measured resistance of $(0.93\pm0.01)\ \Omega$, $(1.93\pm0.02)\ \Omega$, $(1.43\pm0.01)\ \Omega$, respectively. Each of the coils is controlled by a High-Finesse bipolar current source (BCS) with analogue control (±10 V). The supplies maximum outputs are 8A/3V and the current noise is suppressed to $<8\times10^{-5}$ A. This allows for accurate dynamic control of the local fields at the atoms during the experiment.

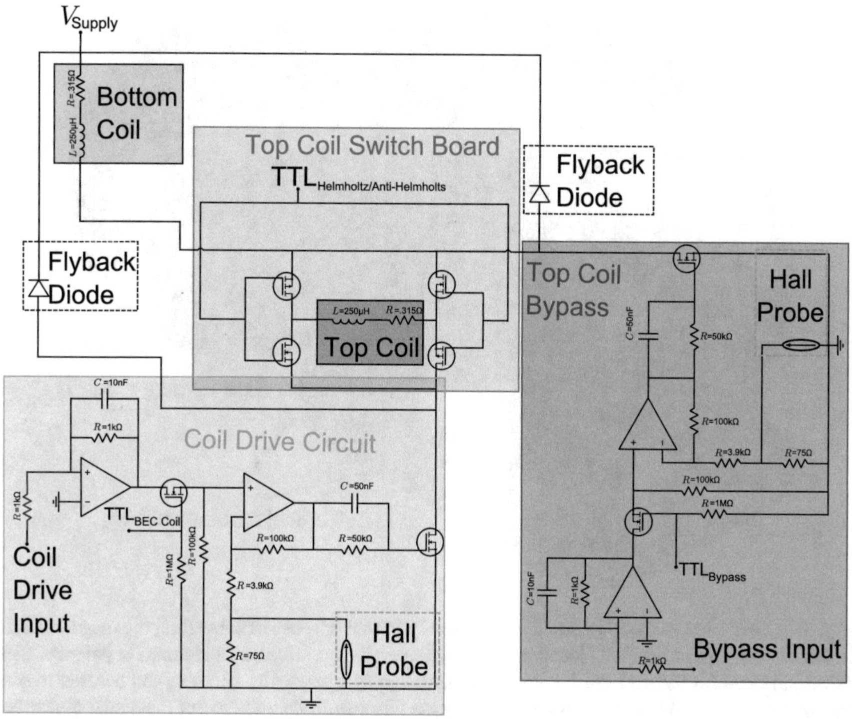

Fig. 3.3 Depiction of the BEC magnetic field coil drive. The consists of a main coil drive circuit which controls the total current going through the coils and monitors that current using a hall probe sensor. The top bypass circuit is used to is used to bypass the top coil so that the current going through the two coils is independent of one another. The top coil is inside a switchboard used to switch the coils from a Helmholtz to an anti-Helmholtz configuration

3.3.1 Levitation Field

To trap atoms with low optical power and levitate the cloud during free expansion, we create a levitation field using our BEC coil pair. This is done by independently controlling the current going through the top and bottom coil. We have two requirements for the levitation field, the first is that the levitation field magnetic gradient in the vertical direction must counteract the effect of gravity on the atoms in the $5^2S_{1/2}|F = 1, m_F = -1\rangle$ state. The second is that the magnetic field gradient perpendicular to gravity should be zero so as to not impede the free expansion of the cloud. To find the currents that satisfy these conditions, we calculate the magnetic field created by the coils. To do so we must first calculate the magnetic field created by one turn of one coil, which is given by

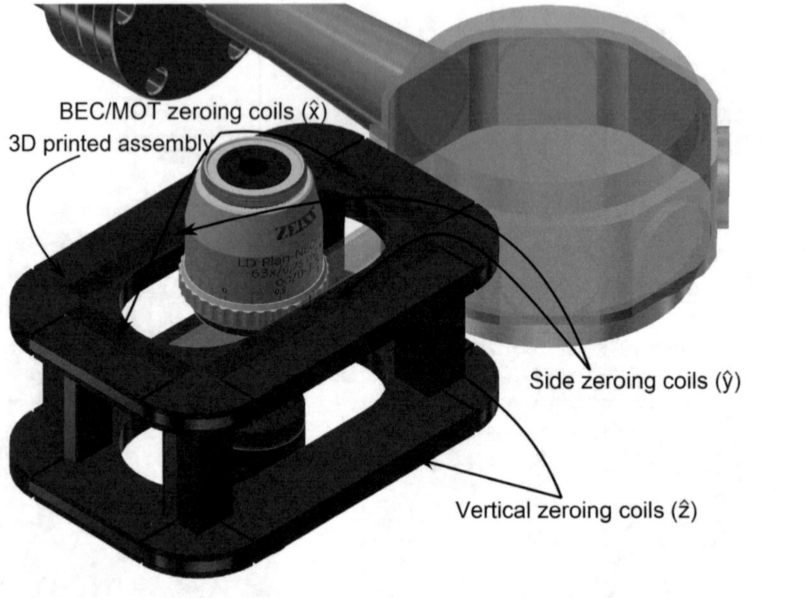

Fig. 3.4 Depiction of the 3D printed coil assembly for the purpose of cancelling the magnetic field in the science cell at the BEC location. The vertical coils are driven as a Helmholtz pair and each of the side and MOT/BEC zeroing coils can be driven independently and thus can be used to bias the system. Depicted objectives are for illustration purposes only and are not the actual objectives used

$$\vec{B}_z(\rho, z, I) = \frac{\mu_0 I}{2\pi\sqrt{z^2 + (a+\rho)^2}} \left[\frac{a^2 - z^2 - \rho^2}{z^2 + (\rho - a)^2} E_E(k) + E_K(k) \right] \hat{z}, \tag{3.1a}$$

$$\vec{B}_r(\rho, z, I) = \frac{\mu_0 z I}{2\pi\rho\sqrt{z^2 + (a+\rho)^2}} \left[\frac{a^2 + z^2 + \rho^2}{z^2 + (\rho - a)^2} E_E(k) + E_K(k) \right] \hat{\rho}, \tag{3.1b}$$

where $k = 4a\rho/(z^2 + (a+\rho)^2)$, E_E (E_K) are complete elliptical integrals of the second (first) kind, a is the radius of the coil, z is the vertical displacement from the center , ρ the radial displacement, μ_0 is the vacuum permeability and I is the current through the coils. The combined magnetic field due to the one turn of one coil is given by $\vec{B}_{coil}(\rho, z, I) = \vec{B}_z(\rho, z, I) + \vec{B}_r(\rho, z, I)$. The combined magnetic field of both coils is then given by

$$\vec{B}(z', \rho) = N_{turns} \left[\vec{B}_{coil}(\rho, z' + d/2, I_{bottom}) + \vec{B}_{coil}(\rho, z' - d/2, I_{top}) \right], \tag{3.2}$$

where d is the separation of the coils, z' is the displacement from the center of the two coils, N_{turns} is the number of turns wound around each coil, $I_{top}(I_{bottom})$ are the current through the top(bottom) coil. For our BEC coils assembly the parameters are a = 50 mm, b = 79 mm, and $N_{turns} = 58$.

To find the condition where the levitation field counteracts gravity we need to know what force the atoms experience in a magnetic field which is due to the magnetic Zeeman shift of the state, leading to an effective potential due to the change in the energy of the state with changes in the magnetic field. The force due to a magnetic field gradient is therefore given by

$$\vec{F}(\vec{r}) = -\vec{\nabla} \cdot \Delta E_{\text{Zeeman}}(\vec{r}) = -\mu_B m_F g_F \vec{\nabla} \cdot \left|\vec{B}(\vec{r})\right|. \tag{3.3}$$

With this we can now solve the equation for the coil currents that leads to our stipulated conditions mainly being

$$\vec{F}(\vec{r}) = g\hat{z}, \tag{3.4a}$$

$$\frac{d\left|\vec{B}(\vec{r})\right|}{d\rho} = 0. \tag{3.4b}$$

Solving for the currents that satisfy Eq. (3.4b), we get that $I_{top} = -0.119 I_{bottom}$ meaning that we still need the coils to be in an anti-Helmholtz configuration but the current of the top and bottom coils need to be controlled independently, which is done with the circuit described in Fig. 3.3. Using the condition for a uniform radial field, solving Eq. (3.4a) indicates that the current through the bottom coil needs to be 26.44 A. Figure 3.5 shows the magnetic field magnitude for horizontal ($\hat{\rho}$) and vertical ($\hat{z}$) displacements from the atom cloud position. The magnetic field magnitude is mostly independent from horizontal displacements as can be seen in Fig. 3.5A and has a linear slope of −30.5 G/cm for horizontal displacement shown in Fig. 3.5B.

3.4 Laser System

In this section, the optical systems that have been modified or added during the tenure of my PhD are described. Most of the optics drawn in this section have used the ComponentLibrary, a library of standard optical components, as their basis, which was created by Alexander Franzen [3].

The laser distribution board to distribute the light to the atoms is presented, as well as the optical beams projected onto the atoms horizontally and vertically through the high resolution objectives which are used to create the external atom potentials for our experiments. The horizontal system is mainly used for 3DMOT cooling, horizontal time of flight imaging and optical trapping in an ODT and in a light sheet. The vertical system which is used for vertical in situ and time-of-flight (TOF) imaging, horizontal

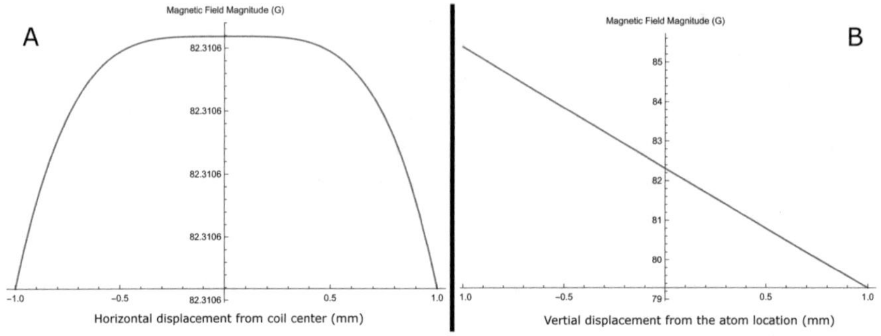

Fig. 3.5 Modelled levitation field for current of bottom(top) BEC coils set to I(−0.119I) where I is set to 26.44 A such that the vertical gradient is −30.5 G/cm to counteract gravity. **A** shows the generated magnetic field magnitude as a function of horizontal displacement from the atom cloud position and it can be seen that the gradient is close to 0 over 2 mm. **B** is the same as in **A** but for vertical displacements from the atom cloud position. It shows an almost uniform gradient of −30.5 G/cm in the magnetic field over 2 mm counteracting gravity uniformly over the 9 μm in-trap extent of the cloud

trapping though a projection of the DMD trapping potential and for spatially selective atom scattering/cloud heating.

3.4.1 Distribution Boards

Figure 3.6 shows the distribution boards of the ^{87}Rb cooling and repump lasers (^{41}K lasers are not shown see [1] for further details). There are two inputs to the system: the repump laser input locked to the $5^2S_{1/2}|F = 1\rangle \rightarrow 5^2P_{3/2}|F' = 1, 2\rangle$, and the seed cooling laser input locked to the $5^2S_{1/2}|F = 2\rangle \rightarrow 5^2P_{3/2}|F' = 2, 3\rangle$ where the $|F' = i, j\rangle$ denotes the crossover line between hyperfine states i and j. The cooling light is up-shifted by a double-pass acousto-optic modulator (AOM) which is used to control the detuning of the cooling light. It is then amplified by a DLX110 tapered amplifier. Some of the light goes into the Bragg set-up (shown in Fig. 3.8), and the rest is down-shifted by a single pass AOM at a fixed frequency and the distributed to the push beam, 2DMOT beams, and 3DMOT distribution board. The repump beam is upshifted by a single pass AOM and distributed to the horizontal 2DMOT, vertical repumping beam, the Faraday imaging board (see Fig. 3.9), and the 3DMOT distribution board. On the 3DMOT distribution board, the repump and cooling are combined and the distributed to six 3DMOT beams and the horizontal imaging beam. The Faraday light and imaging light are also combined and coupled into the vertical imaging fibre. The seed cooling laser input to the distribution board is locked to the $5^2S_{1/2}|F = 2\rangle \rightarrow 5^2P_{3/2}|F' = 2, 3\rangle$ cross-over resonance using standard saturated absorption spectroscopy, as shown in Fig. 3.7.

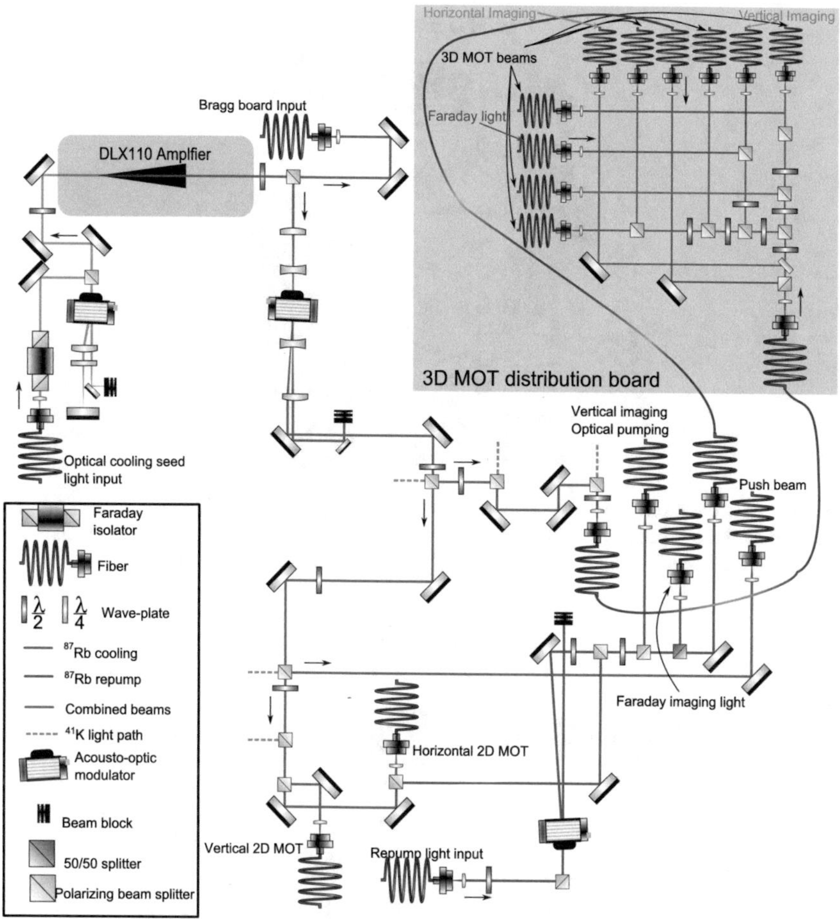

Fig. 3.6 Laser main distribution board layout for ^{87}Rb with 3DMOT distribution board shown as an inset. The board takes in a repump and a cooling seed light and feeds the Bragg distribution board, 2DMOT and 3DMOT, horizontal and vertical absorption imaging, and the Faraday imaging. The green dashed lines represents paths that are used by the ^{41}K light for which the optics are not depicted

The Bragg distribution board depicted in Fig. 3.8 takes in light from the cooling path on the distribution board, which is 80 MHz detuned, from the $5^2S_{1/2}|F = 2\rangle \rightarrow 5^2P_{3/2}|F' = 3\rangle$ transition and splits the light beam into S and P-polarized light with equal power in each arm. For a description of Bragg scattering see Sect. 2.5.4. Each arm is then sent through a double pass AOM with central double pass frequency of -200 MHz bringing the light close to the $5^2S_{1/2}|F = 2\rangle \rightarrow 5^2P_{3/2}|F' = 2, 3\rangle$ cross-over transition with a tunable offset between the 2 beams for momentum selection. After recombining the beams on a polarization beam splitter (PBS), the light is sent to the atom through a polarization maintaining (PM) fibre.

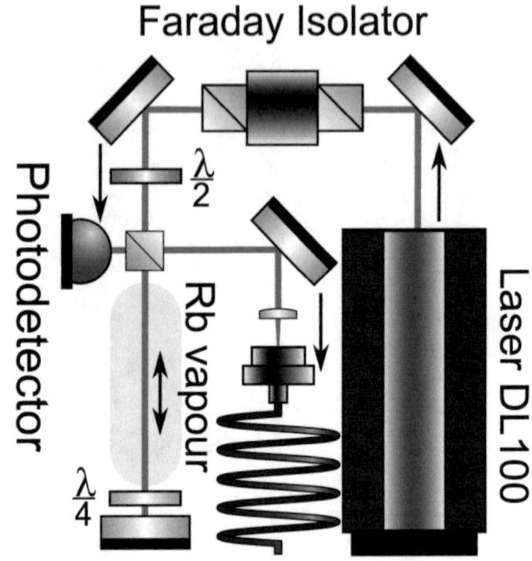

Fig. 3.7 Schematics of the ^{87}Rb master oscillator for the cooling light, consisting of an external cavity diode laser (Toptica DL100) with a simple saturated absorption spectroscopy scheme. This laser is locked to the $5^2S_{1/2}|F=2\rangle \rightarrow 5^2P_{3/2}|F'=2,3\rangle$ cross-over resonance

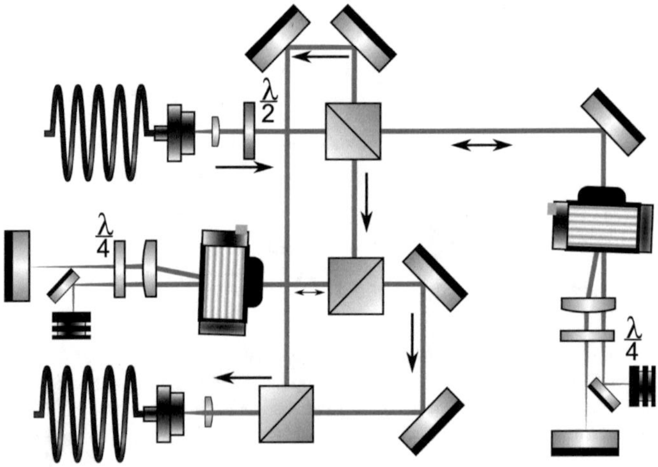

Fig. 3.8 Schematics of the double Bragg beam distribution board which outputs light of two opposite linear polarizations with a slight frequency offset between the two polarizations ($\sim$15.08 kHz). The input light is linearly polarized and +80 MHz from the $5^2S_{1/2}|F=2\rangle \rightarrow 5^2P_{3/2}|F'=3\rangle$ transition the beams is rotated by a wave-plate and both polarizations are sent to double-pass AOMs centred about −100 MHz, with slight offset frequency between the 2 AOMs, bringing the light to the $5^2S_{1/2}|F=2\rangle \rightarrow 5^2P_{3/2}|F'=2,3\rangle$ cross-over transition with a tunable offset between the 2 beams. The light is then recombined and coupled to a fibre, after which it is sent to the BEC to preform Bragg scattering as shown in Fig. 3.11

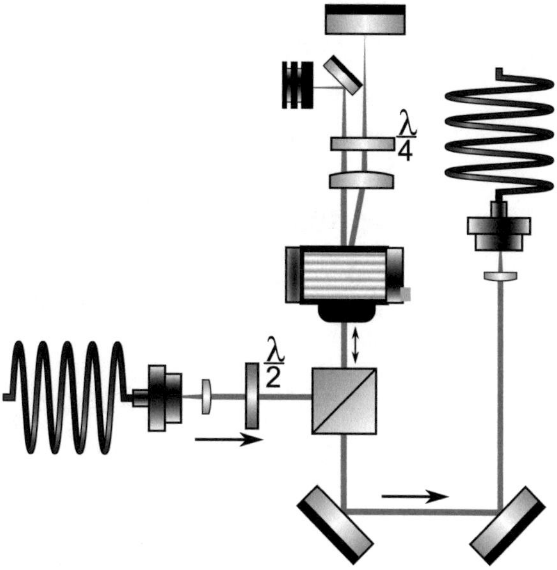

Fig. 3.9 Schematics of the Faraday light distribution board. The input light comes from the repump path shown in Fig. 3.11 and is on resonance with the $5^2S_{1/2}|F = 1\rangle \rightarrow 5^2P_{3/2}|F' = 2\rangle$ transition. The light is then passed through a double-pass AOM to tunably shift the frequency, usually set to 95 MHz (−190 MHz total shift), this light is then coupled to a fibre and sent to the distribution board shown in Fig. 3.11

The Faraday light distribution board (Fig. 3.9) takes light on resonance with the $5^2S_{1/2}|F = 1\rangle \rightarrow 5^2P_{3/2}|F' = 2\rangle$ transition and then outputs it with a tunable detuning relative to the input by using a double pass AOM. The detuned light is then coupled to a fibre and sent to the distribution board (Fig. 3.6). On the board, it is coupled with the absorption imaging light into a PM fibre and then sent to the vertical imaging system. Having both lights coupled to the same fibre allows for a quick change between the absorption and Faraday imaging on the vertical system, by simply rotating the output polarization before it goes through a final PBS cube. For the physics of how to Faraday image atoms see Sect. 2.5.3.

To project the DMD pattern onto the atom plane, we use repulsive green light (532 nm) depicted in Fig. 3.10. For the full details of the projection see Chap. 4: here only the physical system will be described. The green light is generated by a Spectra-Physics Millennia laser with output power tunable from 0.2 to 5 W. The green light beam is first expanded by a factor of 2X to prevent issues with thermal deformation of lenses and mirrors when high laser power is used. The light is then split using a PBS cube to be sent either to the optical accordion which we are still working on or sent to a fibre to be used to project our DMD onto the atom plane. The light going to the DMD is magnified by a factor of 1/3, then sent through an AOM driven by multiple frequencies. The green light is then coupled to an Endlessly Single Mode (ESM), large-mode-area, polarization-maintaining, photonic crystal fibre with an NA of 0.048 ± 0.02. Due to the small NA of the fibre, any angular shift of the AOM output beam leads to a significant change in the coupling. We found that thermal drifts of the output angle of the AOM due to heating up of the crystal after turning on the device were detrimental to the reproducibility of our experiments. So following [4], we decided to feed two frequencies to our AOM and maintain the power between the

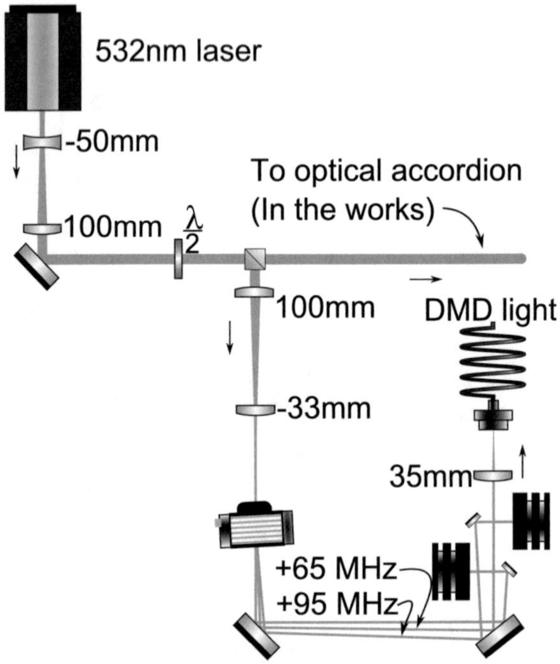

Fig. 3.10 532 nm laser light distribution board. The light from the laser is expanded and split into two paths one for that will be used to supply the optical accordion [5] (still to be implemented in the system) and the rest of the power is coupled to a fibre to be used to illuminate the DMD. It is coupled through an AOM that acts as a switch. Due to the low NA of the 532 nm fibre (0.048 ± 0.02) the drift in the beam angle at the output of the AOM affects the coupling into the optical fibre and causing drifts in the system. To minimize these, we used the scheme proposed by [4] of driving the AOM with two frequencies and making sure that the sum of the power of the acoustic waves was constant (color figure online)

two such that the power coming into the AOM was constant. This way, even when the AOM is not being used to send light to the DMD, the dissipated acoustic power in the crystal is constant which leads to a more stable temperature of the crystal thereby mitigating the thermal angular drift of the AOM.

3.4.2 Laser Cooling, Trapping, Imaging and Bragg Systems

The optical system on the vacuum side is made up of the horizontal (see Fig. 3.11) and vertical (see Fig. 3.12) optical systems.

The horizontal system consists of five main components. The 3DMOT optical system consisting of six 12.7 mm waist optical beams with 8 mW of cooling light and 100 μW repumping light set to circular polarisation. A 9.5 mm waist imaging

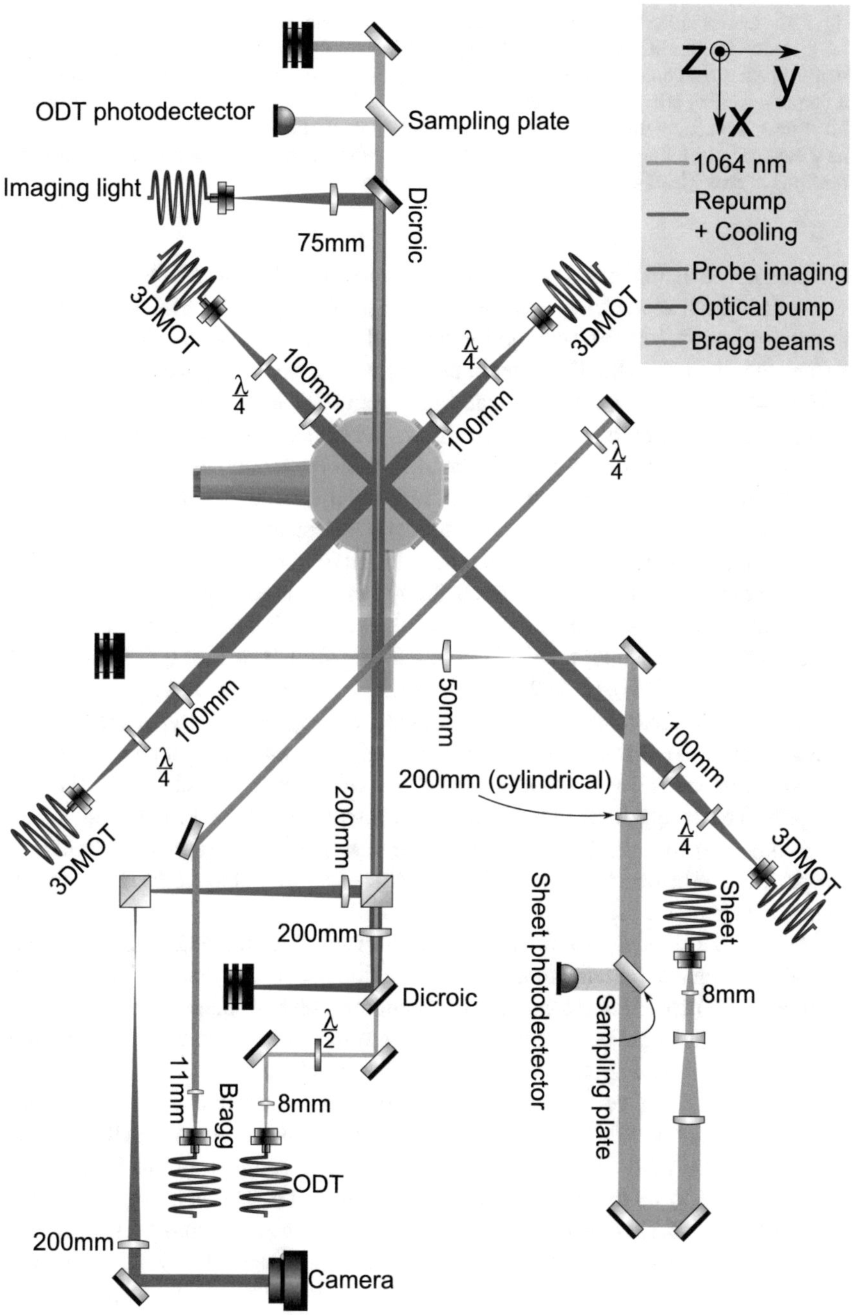
z
y
x
1064 nm
Repump
+ Cooling
Probe imaging
Optical pump
Bragg beams
ODT photodectector
Sampling plate
Imaging light
Dicroic
75mm
3DMOT
3DMOT
100mm
100mm
λ/4
λ/4
λ/4
50mm
100mm
100mm
3DMOT
3DMOT
200mm (cylindrical)
200mm
200mm
Dicroic
λ/2
8mm
11mm
Bragg
ODT
Sheet photodectector
Sampling plate
Sheet
8mm
200mm
Camera

◂**Fig. 3.11** Optics in the horizontal plane near the 3DMOT and Science Cell. The optics consist of five systems: (1) The 3D MOT optics which provides our initial cooling stage made up of 6 beams (vertical beams not depicted); (2) A horizontal imaging system which consists of the imaging light in purple travelling along the $\hat{x}$-direction and a Prosilica GX1050C CCD camera; (3–4) In orange (1064 nm), the optical dipole trap (ODT) travelling through the cell along the $-\hat{x}$-direction, and the sheet beam travelling through the cell along the $-\hat{y}$-direction; (5) A double-Bragg beam set-up travelling at an angle of 45° from the y-axis beam in green

light beam, consisting of 500 μW of linearly P-polarized light and 100 μW of linearly S-polarized light repumping light, is made to pass through the BEC to produce an absorption image. The probing light is sent to the camera through a PBS to separate it from the repump light. The camera used to image the probing light is a Prosilica GX1050C CCD Camera. The optical dipole trap (ODT) is sent to the optical table through a fibre and is focused onto the atoms for optical trapping with 95 μm waist and power tunable between 50 mW and 4 W with power actively stabilized with a photodetector [6]. A sheet beam with stabilized adjustable power between 4.2 mW and 250 mW focused onto the atom cloud with a vertical direction waist of 8.6 μm and collimated horizontal waist of 500 μm is used for the vertical confinement of the atoms in the final BEC. The horizontal Bragg beam is used to probe the momentum of the cloud and measure the circulation of the vortices in the system [7]. The beam is generated on the Bragg distribution board (Fig. 3.8). The Bragg light is 80 MHz detuned from the $5^2S_{1/2}|F = 2\rangle \rightarrow 5^2P_{3/2}|F' = 3\rangle$ transition and so is ~6.8 GHz detuned from the closest transition for atoms in the $5^2S_{1/2}|F = 1\rangle$ hyperfine state. The Bragg beam consists of S and P-polarized light with controllable detuning between the two components for momentum selection, with power variable from 10 μW to 500 μW. The Bragg light is made to overlap with the atoms at a 45° angle with the y-axis, and then is retro-reflected through a quarter wave-plate to switch the polarization of the beams. This results in a lattice to diffract the atoms in the double-Bragg configuration [8].

The vertical optical system (Fig. 3.12) consists of an imaging path that can do both absorption and Faraday imaging, a projection path to project the DMD onto the atom plane, to provide horizontal confinement through blue-detuned repulsive light, and a third path to project a second DMD onto the atom plane with near-resonant light, allowing us to selectively destroy parts of the cloud.

The light for the vertical imaging system comes from the distribution board (Fig. 3.6) and can be either set to be probing light, which combined with the repumping light can be used for absorption imaging or can come from the Faraday distribution board to be directly used for Faraday imaging. The imaging system (see right half of Fig. 3.13) consists of a microscope objective (Olympus LCPLN20XIR) with a 0.45 NA, with a glass aberration correction collar for 0–1.2 mm thickness, antireflection coated for near-infrared radiation, with a 9 mm effective focal length and a 8.93–8.18 mm front working distance (dependent on the glass thickness). The atoms are then re-imaged using a 100 mm achromatic lens which produces an 11.11× magnified image. The image is then projected using a lens relay, which is not used at the moment, but is there for the implementation of a dark field imaging in

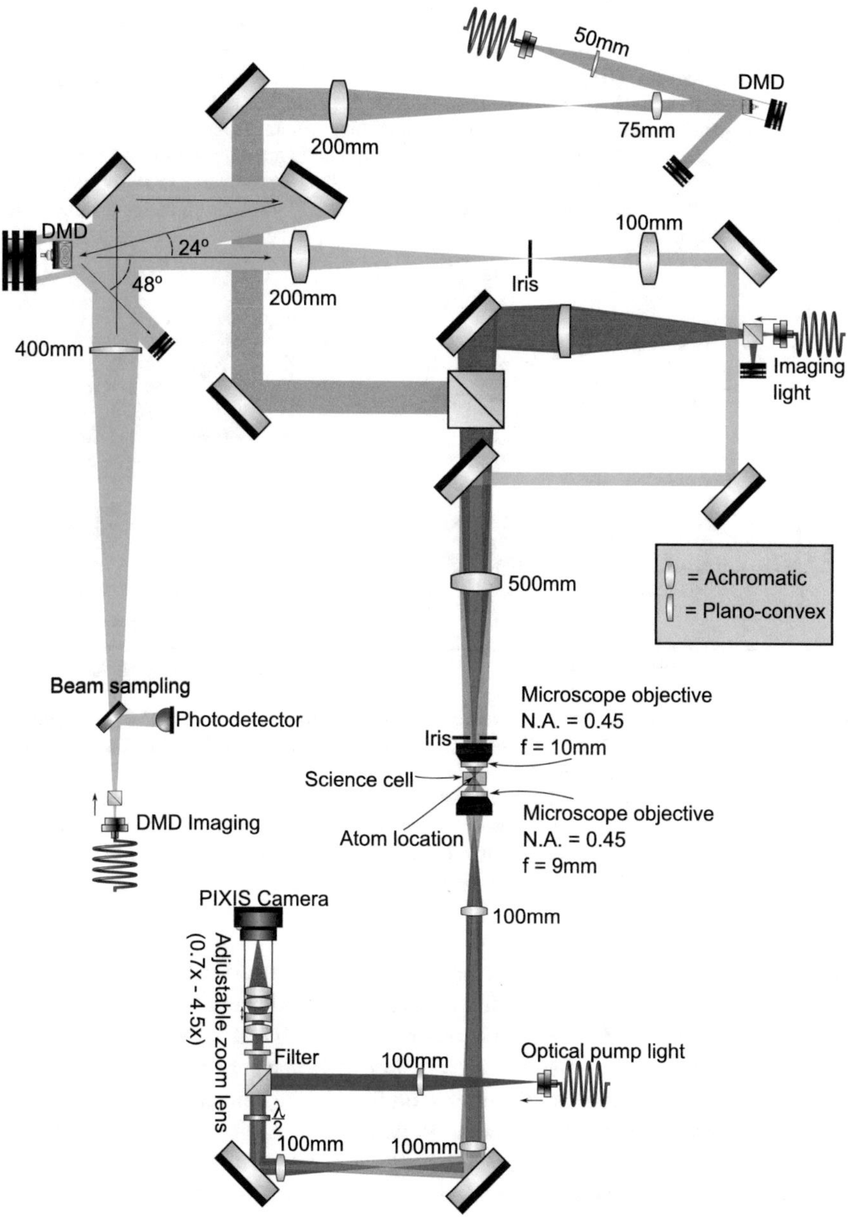

50mm
DMD
200mm
75mm
DMD
24°
48°
100mm
200mm
Iris
400mm
Imaging
light
500mm
= Achromatic
= Plano-convex
Beam sampling
Photodetector
Microscope objective
N.A. = 0.45
f = 10mm
Iris
Science cell
Microscope objective
N.A. = 0.45
f = 9mm
DMD Imaging
Atom location
PIXIS Camera
100mm
Adjustable zoom lens
(0.7x - 4.5x)
Filter
100mm
Optical pump light
λ/2
100mm
100mm

◂**Fig. 3.12** Schematic representation of all the vertically travelling optical systems at the BEC location. The green light is used to project an image of the DMD onto the atom plane through a 100× imaging system. The red light is used to probe the cloud and it can either be tune to be Faraday or absorption light for imaging. The orange light is the 'kill light' which is masked by a DMD and projected onto the light plane to optionally remove atoms from the trap. The blue beam is the repump light used in standard absorption imaging and is the only light travelling upwards through the cell. Note that the adjustable zoom lens is only shown as a schematic the actual internal optics are unknown. The zoom lens model we use is a VZM^{TM} 450i Zoom Imaging Lens with a 90 mm working distance

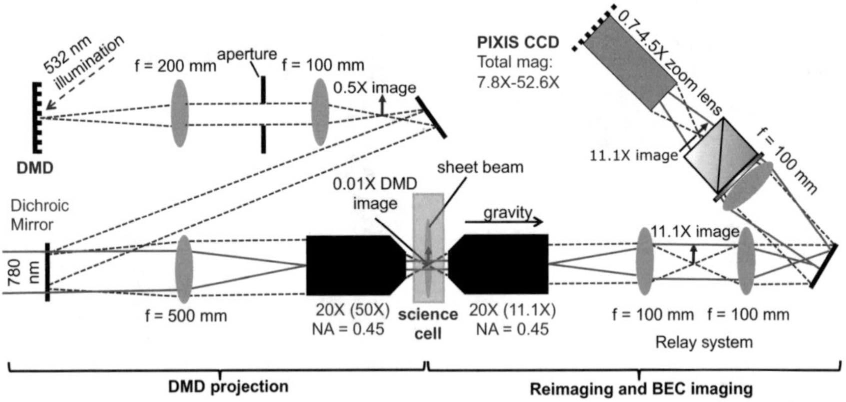

Fig. 3.13 DMD projection system and reimaging/BEC imaging system. The DMD is imaged on to the atom plane with 0.01× magnification (100× minification), and combined with a red-detuned TEM_{00} sheet to form configurable 2D trapping potentials. The bottom system similarly produces an image of the BEC or DMD pattern, with adjustable magnification from 7.77×52.6×. The half wave plate is used to switch between Faraday imaging and standard absorption imaging. The lens relay system will be used for a future implementation of darkground imaging [9, 10]. Note this figure is an adapted from our published paper [11]

the future [9, 10]. The intermediate image is then re-imaged onto the CCD camera (PIXIS 1024B) using a fixed working distance VZM450 zoom lens (0.70×–4.74×) allowing for 7.77×–52.6× adjustable magnification of the atom cloud onto the camera. The beam cube and half-wave plate are used to switch between Faraday imaging, where the probe light does not illuminate on the camera, and absorption, where the probe light does make it to the camera. As part of the imaging system, a repumping beam is also sent backwards through the imaging system to repump the atoms before they are probed. This path is only used during vertical absorption imaging, as the Faraday detection works on the repump absorption lines.

The projection of the DMD used for horizontal confinement is done using 532 nm light coming from the green laser light distribution board. The light power is first stabilized and monitored using a photodetector and PID controller circuit feeding back onto the AOM signal on the distribution board. The illumination light of Gaussian shape, with $w_0 = 9.81$ mm, is projected onto the DMD (WUXGA 1200×1920 pixels with Visitech LUXBEAM 4600 controller) with an incident angle of 24°. The reflected light by the 'on' mirrors is propagated through the imaging system to the atoms while the light reflected by the 'off' mirrors is absorbed by a beam block. For further information about the operation of the DMD see Sect. 4.3. The measured diffraction efficiency is $\sim$40%. The image is projected using a 1/2$\times$ system composed of visible wavelengths AR-coated achromatic lenses of 200 mm and 100 mm focal lengths, respectively. An iris is inserted in the Fourier plane of the 200 mm lens to filter out the unwanted extra diffracted orders of the DMD, which otherwise contribute stray light to the dark regions of the imaged pattern reducing the uniformity of the trapping potential and the lifetime of the atoms trapped in the dark regions. The 1/2$\times$ projected image of the DMD is then re-imaged with a 1/50$\times$ imaging system consisting of a 500 mm achromatic lens and Nikon CFI PLAN EPI 20XCR infinity corrected objective also with 0.45 NA, glass aberration correction collar, visible AR-coating, 10 mm effective focal length, and 10.9–10 mm working distance. This objective is made for inspecting LCD panels which is the same thickness as our science cell. Overall the DMD pattern is projected onto the atom plane with a magnification of 1/100$\times$. The quality of the projected pattern was found to be insensitive to the angle between the microscope objective plane and the atom imaging plane up to coarse adjustment, but we found that correction for the glass aberrations was very important.

The third vertical optics system projects another DMD to create a spatially patterned atom cloud "kill" beam. It consists of light tuned to $5^2S_{1/2}|F = 1\rangle \rightarrow 5^2P_{3/2}|F' = 2\rangle$ reflected off a DMD (DLP LightCrafter). The DMD is first imaged using a 8/3$\times$ system made up of a 75 mm and 200 mm achromatic lenses. The light is then projected onto the atom plane using the same 1/50$\times$ imaging system as for the horizontal trapping projection pattern resulting in a 4/75$\times$ magnification of the original DMD when projected onto the atom plane.

3.5 BEC Creation Summary

In this section, the experimental steps used to create a ^{87}Rb Bose-Einstein condensate in the $5^2S_{1/2}|F = 1, m_F = -1\rangle$ state will be described. This state can be manipulated using either light due to the dipole interaction or a magnetic field since the state is in a low magnetic field seeking state.

The steps described in detail in this section are first the cooling step which uses a 2D Magnetic Optical Trapping (MOT) combined with a 3DMOT to collect, trap and cool atoms down to the μK regime through photon recoil cooling. Then, the atoms are pumped into the $5^2S_{1/2}|F = 1, m_F = -1\rangle$ trappable state and trapped in a quadrupole magnetic field, then magnetically transferred from the octagon area to the Science Cell area (see Fig. 3.1). Next, in a high quadrupole field gradient, microwave evaporation is used to further cool the cloud. Then, the atoms are loaded into an optical dipole trap combined with a magnetic field in which optical evaporation is performed. Finally, the atoms are transferred from the optical dipole trap into a sheet trap where optical evaporation is further performed. At the end of the evaporation ramp, before the atom condense, the DMD projected pattern is turned on such that the atoms condense inside the DMD pattern.

3.5.1 Magnetic Optical Trapping and Cooling

The principle behind MOT cooling is that an atom put inside an optical beam will experience an optical force due to the directionality of the incident photons and the isotropy of the reemission. This scattering will lead to an effective friction force which will slow down the atoms interacting with the light thereby cooling them [12]. Using either 6 beams of light coming along each coordinate axis is not enough to produce a trap since random walk could allow the atoms to stray from the trap (Fig. 3.11). To remedy this the extra force is made to be spatially dependent to provide the restoring force needed to create a trap. This is done by applying a magnetic gradient field at the locations of the atoms which introduces spatially dependent Zeeman shifts of the internal energy levels of the atoms [13].

For the cooling of our ^{87}Rb atoms, we use the $D2$ transitions lines (see Fig. 2.2) with an optical cooling laser driving the transition $5^2S_{1/2}|F=2\rangle \rightarrow 5^2P_{3/2}|F'=3\rangle$ with circularly polarized $\sigma_\pm$ which forms a closed transition between $5^2S_{1/2}|F=2, m_F=\pm 2\rangle \rightarrow 5^2P_{3/2}|F'=3, m_{F'}=\pm 3\rangle$. Not all of the atoms start in the $5^2S_{1/2}|F=2\rangle$ hyperfine state, as some start in the $5^2S_{1/2}|F=1\rangle$ state, and it is possible to lose atoms from the cyclical transition due to spontaneous emission down to the $5^2S_{1/2}|F=1\rangle$ state. To prevent these losses, a repump laser that drives the atoms from $5^2S_{1/2}|F=1\rangle \rightarrow 5^2P_{3/2}|F'=2\rangle$ is used. The atoms through spontaneous emission will decay back either into $5^2S_{1/2}|F=1\rangle$ state which will be re-pumped again or $5^2S_{1/2}|F=2\rangle$ state which is back in the cooling cycle. This effectively creates a closed transition cycle for our cooling.

We use a two-stage loading process where on the low vacuum side (due to high ^{87}Rb partial pressure) of our experiment (see Fig. 3.1) the atoms are cooled down in two directions in the 2DMOT. The low vacuum allows for a high atom density but is detrimental to the lifetime of a BEC when doing experiments. Therefore, the atoms are pushed to the ultra-high vacuum side of the experiment, using a push beam along the axis of the 2DMOT cloud that is not cooled, through a pressure differential tube of 12 mm of length and 1.2 mm inner diameter providing the low conductance needed to maintain the pressure differential (~15,000) between the two vacuum chambers. The atoms are then transferred to a 3DMOT where they are cooled along all three directions while sitting inside a quadrupole field generated by a pair of anti-Helmholtz coils.

The 2DMOT and 3DMOT repump laser are tuned to the $5^2S_{1/2}|F=1\rangle \rightarrow 5^2P_{3/2}|F'=2\rangle$ transition and the cooling lasers are red detuned by $-3\Gamma_{D2}$, where $\Gamma_{D2}=2\pi\times 6.065$ MHz is the linewidth of the D2 transition [14], for the $5^2S_{1/2}|F=2\rangle \rightarrow 5^2P_{3/2}|F'=3\rangle$ transition. The vacuum pressure on the low vacuum side is about $\sim 10^{-7}$ Torr and about $\sim 5\times 10^{-12}$ Torr on the ultra-high vacuum side. We load $\sim 10^{10}$ atoms in our 3DMOT in about 10 s with a temperature of around 100 μK.

3.5.2 Compressed MOT and Repumping

The density of atoms remains quite low in the 3DMOT due to the scattering of light inside the cloud. To avoid heating of the magnetically trappable atoms due to energy gained during the increase in the magnetic field, the density of the atomic cloud is increased. This is done by performing a compressed MOT (CMOT) [15]. This step involves linearly ramping the cooling laser detuning from $-3\Gamma \rightarrow -10\Gamma$ over 50 ms and leaving the magnetic field gradient at 10.5 G/cm while reducing the repump power to about 10% of its initial value. This allows the density to grow due to the reduce scattering rate coming from the centre of the cloud leading to a reduced internal pressure of the cloud allowing the density to increase.

Most of the atoms are in the $5^2S_{1/2}|F = 1\rangle$ manifold due to our low repump power during the CMOT step, but some atoms are still in the $5^2S_{1/2}|F = 2\rangle$. We want our atoms in the $5^2S_{1/2}|F = 1, m_F = -1\rangle$ state, so the repump beam is turned off for 2 ms after the CMOT to transfer all of the atoms into the $5^2S_{1/2}|F = 1\rangle$ manifold.

3.5.3 Magnetic Trapping and Transfer

We slightly favour the magnetically trappable $5^2S_{1/2}|F = 1, m_F = -1\rangle$ state by leaving the magnetic field on during the cooling and depumping process but still many of the atoms end up in non-magnetically trappable states $5^2S_{1/2}|F = 1, m_F = 0, 1\rangle$. This means that when ramping on our magnetic field gradient to 100 G/cm over 100 ms, we trap only a fraction of the atoms $1/3 < \eta < 1/2$ which was measured to be ∼45%. This magnetic trapping potential is used in concert with two additional anti-Helmholtz coil pairs [2] (see Fig. 3.2),to transfer the atom cloud from the 3DMOT to the Science Cell where further cooling will take place. Even though no sub-Doppler cooling is performed [16], the measured temperature of the magnetically captured cloud generally is at around 80 μK with an atom number of $2 - 3 \times 10^9$.

The atoms are transferred over 800 ms from the Science Cell with a Hanning window velocity and acceleration profile [17]. This process is fairly adiabatic with no measurable atom loss, but increases the temperature of the cloud to about 100 μK. The quadrupole field is subsequently increased to 145 G/cm over 750 ms, prior to the initial microwave evaporation.

3.5.4 Microwave Evaporation and Optical Dipole Trap Transfer

To further cool down our cloud in the magnetic trap we use energy selective microwave (μ-wave/MW) evaporation [18, 19]. Evaporation occurs over 4 s by driving the μ-wave transition form the magnetically trappable $5^2S_{1/2}|F = 1, m_F = -1\rangle$

state to the magnetically untrappable $5^2S_{1/2}|F = 2, m_F = -1\rangle$ state. At the end of the process, we obtain an atomic cloud with $1.5 - 2 \times 10^8$ atoms at a temperature of 15 μK.

We cannot reach the BEC transition using MW-evaporation due to the Majorana loss processes [20] that happen at the centre of the trap, where the magnetic field goes to zero and the quantization becomes undefined. This occurs because the frequency at which the atoms spin precess (Larmor frequency) is so low that the atom spins cannot follow the rapid change in the magnetic field direction, causing spin-flips of the $5^2S_{1/2}|F = 1, m_F = -1\rangle$ atoms to untrappable states such as $5^2S_{1/2}|F = 1, m_F = 0, 1\rangle$ [20]. This leads to Majorana losses of atoms which will scale up as the density of atoms near the centre of the trap decreases as the density and temperature of atoms drops during the evaporation process preventing the system from ever reaching the macroscopic ground state [21].

There are many solutions to circumvent this problem. One is to prevent the atoms from reaching the magnetic zero using a repulsive beam that effectively plug the hole in the trap [22]. Another method utilises the Ioffe-Pritchard magnetic trap configuration to make the atoms experience a symmetric trap with a magnetic field offset [23, 24]. In our system, we transfer our atoms to a hybrid magnetic and optical dipole trap after the MW evaporation to avoid this issue, which also allows for further cooling of the atoms [25]. To transfer from the magnetic trap to the optical dipole trap, the quadrupole field gradient is ramped down from 145 G/cm to 27 G/cm over 4 s, while a red-detuned 1064 nm optical dipole trap (ODT) located ~80 μm below the zero of the quadrupole magnetic field with a waist of 49 μm is turned on at full power equivalent to a potential depth of 110 μK × k_B. To prevent the location of the magnetic field from drifting during the ramp down of the magnetic field gradient, we implemented zeroing coils mounted on a 3D printed frame to cancel the background stray magnetic fields (see Fig. 3.4). Due to gravity, once the magnetic field gradient is low enough, the atoms will no longer be trapped in the vertical direction, causing them to slowly transfer to the optical dipole trap with minimal heating. We measure loaded atom numbers of $3 - 4.5 \times 10^7$ atoms at a temperature of 4.5 μK. The temperature is lower than that of the cloud after μ-wave evaporation because only a portion of the atoms are transferred that tend to be the lower energy atoms which disperse less than the higher energy ones during the transfer. We then perform optical evaporation to further lower the temperature of our cloud.

3.5.5 Optical Evaporation and Transfer

The principle behind optical evaporation is simply that atoms with high energy can escape whereas the low energy ones cannot. If one slowly lowers the trap height then the mean kinetic energy per particle in the cloud will decrease thereby decreasing the temperature of the cloud.

We perform optical evaporation by lowering the optical power of the dipole-magnetic hybrid trap from 4.5 W to 400 mW to cool the atoms down to ~450 nK, which is just above the critical temperature of the condensate of about 300 nK. We do not condense the atoms at this step, instead condensing them in the sheet with the digital micromirror device pattern on so as to avoid potential excitations during the transfer due to the mismatch between the sheet and the optical dipole trap frequencies. If we were to continue the optical evaporation further we obtain BECs of $3 - 5 \times 10^6$ atoms in this trap. Instead, we load a cold thermal atoms cloud from the optical dipole trap into a red-detuned 1064 nm beam focused in the vertical direction and collimated in the horizontal direction (i.e. a sheet trap). The beam travels along the $-\hat{y}$-direction, perpendicular to the optical dipole trap travel direction ($\hat{x}$-direction), and has a waist of $(w_x^0, w_z^0) \approx (500\ \mu\text{m}, 8.5\ \mu\text{m})$ and a trap depth of $6.4\ \mu\text{K} \times k_B$ (see Fig. 3.11). The transfer is done by ramping on the sheet trap while ramping off the optical dipole trap and magnetic trap. We then evaporate in the combined sheet-gravity trap by once again lowering the optical power, from ~250 mW down to ~55 mW. Before the atoms start to condense, we ramp on the DMD potential (see Chap. 4 for more information on DMDs) to provide the horizontal confinement. The result is an atomic cloud of $\sim 3 - 4 \times 10^6$ atoms and >85% condensate fraction in a potential radially dictated by the DMD, with inside a sheet-magnetic trap with final trap frequencies $(\omega_{x,y}, \omega_z) \approx 2\pi \times (20{,}310)$ Hz.

3.5.6 *Combined DMD and Sheet and Levitation Potential*

Once the atoms have been condensed, a levitation field is ramped on allowing us to hold on to our atoms in a reduced sheet-DMD trap without losing them due to gravity. The levitation field is generated using the BEC quadrupole coils pair with the driving circuit shown in Fig. 3.3 with the field description shown in Sect. 3.3.1. This leads to a trap with measured trapping frequencies of $(\omega_{x,y}, \omega_z) \approx 2\pi \times (6{,}320)$ Hz with a depth of 800 nK$\times k_B$. This is the trap used for the superfluid transport study Chap. 5. We can lower the power of the sheet even further to obtain a flatter potential with trapping frequencies of $(\omega_{x,y}, \omega_z) \approx 2\pi \times (2{,}105)$ Hz with a potential depth of 84 nK$\times k_B$. This is the trap used to study turbulence as presented in Chap. 6.

3.6 Conclusion

Our ^{87}Rb BEC apparatus is optimized to produce BECs with atom number of $3 - 4 \times 10^6$ with light scattering limited lifetime measured of ~35 s, trapped inside versatile 2D potentials resulting from a horizontally trapping red-detuned sheet trap crossed with a blue-detuned projected digital micromirror device (DMD). The superfluid transport study (Chap. 5) uses the versatility of the projected potential to tune the atomtronic circuit of interest and study how the various quantities of interest change with system geometry. Using the dynamics capabilities of the potential, we can generate turbulence and arbitrary potentials as described in Chap. 4. We can also use our zeroing coils as push coils to change the equilibrium position of our BEC or correct for density gradients. This system thus presents a versatile platform that enables the experiments presented here and will continue to be used for a diverse set of experiments in the lab moving forwards (see Chap. 7).

References

1. Parry NM (2015) Design, construction, and performance towards a versatile 87Rb & 41K BEC apparatus. The University of Queensland
2. Parry NM et al (2014) Note: high turn density magnetic coils with improved low pressure water cooling for use in atom optics. Rev Sci Instrum 85:086103
3. Frazen A (2006) Component library
4. Fröhlich B et al (2007) Two-frequency acousto-optic modulator driver to improve the beam pointing stability during intensity ramps. Rev Sci Instrum 78:043101
5. Ville JL et al (2017) Loading and compression of a single two-dimensional Bose gas in an optical accordion. Phys Rev A 95:013632
6. Meyer K (2010) An optical dipole trap for a two-species quantum degenerate gas English. Diploma thesis. University of Heidelberg, Heidelberg, Germany
7. Kwon WJ, Moon G, Choi J-Y, Seo SW, Shin Y-I (2014) Relaxation of superfluid turbulence in highly oblate Bose-Einstein condensates. Phys Rev A 90:063627
8. Moon G, Kwon WJ, Lee H, Shin Y-I (2015) Thermal friction on quantum vortices in a Bose-Einstein condensate. Phys Rev A 92:051601
9. Wilson KE, Newman ZL, Lowney JD, Anderson BP (2015) In situ imaging of vortices in Bose-Einstein condensates. Phys Rev A 91:023621
10. Pappa M et al (2011) Ultra-sensitive atom imaging for matter-wave optics. New J Phys 13:115012
11. Gauthier G et al (2016) Direct imaging of a digital-micromirror device for configurable microscopic optical potentials. Optica 3:1136–1143
12. Wallis H (1995) Quantum theory of atomic motion in laser light. Phys Rep 255:203–287
13. Pritchard DE, Raab EL, Bagnato V, Wieman CE, Watts RN (1986) Light traps using spontaneous forces. Phys Rev Lett 57:310–313
14. Steck DA (2015) Rubidium 87 D line data 2.1.5
15. Ensher JR (1998) The first experiments with Bose-Einstein condensation of 87Rb MA thesis. University of Colorado Boulder
16. Dalibard J, Cohen-Tannoudji C (1989) Laser cooling below the Doppler limit by polarization gradients: simple theoretical models. J Opt Soc Am B 6:2023–2045
17. Greiner M, Bloch I, Hänsch TW, Esslinger T (2001) Magnetic transport of trapped cold atoms over a large distance. Phys Rev A 63:031401
18. Metcalf HJ, Straten Pvd (2003) Laser cooling and trapping of atoms. J Opt Soc Am B 20:887–908
19. Ketterle W, Druten NJV (1996) In: Bederson B, Walther H (eds) Advance in atomic, molecular, and optical physics, pp 181–236
20. Majorana E (1932) Atomi orientati in campo magnetico variabile. Il Nuovo Cimento 1924–1942(9):43–50
21. Petrich W, Anderson MH, Ensher JR, Cornell EA (1995) Stable, tightly confining magnetic trap for evaporative cooling of neutral atoms. Phys Rev Lett 74:3352–3355
22. Davis KB et al (1995) Bose-Einstein condensation in a gas of sodium atoms. Phys Rev Lett 75:3969–3973
23. Pritchard DE (1983) Cooling neutral atoms in a magnetic trap for precision spectroscopy. Phys Rev Lett 51:1336–1339

24. Bergeman T, Erez G, Metcalf HJ (1987) Magnetostatic trapping fields for neutral atoms. Phys Rev A 35:1535–1546
25. Lin Y-J, Perry A, Compton R, Spielman I, Porto J (2009) Rapid production of $^{87\$}$Rb Bose-Einstein condensates in a combined magnetic and optical potential. Phys Rev A 79:063631

Chapter 4
Configuring BECs with Digital Micromirror Devices

4.1 Introduction

Programmable spatial light modulators (SLMs) have significantly advanced the configurable optical trapping of particles. Often, these devices are utilized in the Fourier plane of an optical system, but direct imaging of an amplitude pattern can potentially result in increased simplicity and computational speed. In this chapter, we demonstrate high-resolution direct imaging of a digital micromirror device (DMD) at high numerical apertures (NA), which is applied to the optical trapping of a Bose-Einstein condensate (BEC). We utilise a (1200×1920) pixel DMD and commercially available 0.45 NA microscope objectives, finding that atoms confined in a hybrid optical/magnetic or all-optical potential can be patterned using repulsive blue-detuned (532 nm) light with 630(10) nm full-width at half-maximum (FWHM) resolution, which is within 5% of the diffraction limit. The result is near arbitrary control of the density the BEC without the need for expensive custom optics. We also introduce the technique of time-averaged DMD potentials, demonstrating the ability to produce multiple grayscale levels with minimal heating of the atomic cloud, by utilising the high switching speed (20 kHz maximum) of the DMD. The quality of these time-averaged DMD potentials can be improved by feedback on the atomic density allowing for the realization of truly beyond binary potentials. These techniques have enabled the realization and control of diverse optical potentials for the study of superfluid transport dynamics (Chap. 5) and the realization of Onsager vortices (Chap. 6) with our quantum gases.

Part of the work presented in this chapter has been published by Science in the following publication: **G. Gauthier**, I. Lenton, N. McKay Parry, M. Baker, M. Davis, H. Rubinsztein-Dunlop and T. W. Neely, Direct imaging of a digital-micromirror device for configurable microscopic optical potentials, *Optica* **3**, 1136–1143 (2016).

© The Editor(s) (if applicable) and The Author(s), under exclusive license to Springer Nature Switzerland AG 2020

G. Guillaume, *Transport and Turbulence in Quasi-Uniform and Versatile Bose-Einstein Condensates*, Springer Theses,
https://doi.org/10.1007/978-3-030-54967-1_4

4.2 Background

The manipulation of microscopic particles has benefited from the high level of control and measurement provided by optical tweezers. With the technological development of fast configurable spatial light modulators (SLMs) allowing for ever more complex trapping geometries [1–3], new applications have emerged. For example, sculpted light may have an important role in overcoming multiple light scattering issues in complex biological tissues, and such biomedical applications have only begun to be explored. In particular, the development of sculpted light patterns across the image plane, such as the generation of large trapping arrays, could have application to the in vivo trapping of larger objects, such as living cells [4].

In degenerate quantum gases, the push for increased diversification of optical trapping potentials has led to the adoption of many of the techniques from holographic optical tweezers. SLMs are most often used in the Fourier plane of an optical system, manipulating the phase of an input optical field to produce a configurable pattern in the conjugate trapping plane of the system [5–8]. These methods have been successfully applied to address and pattern atoms trapped in optical lattices [9–11], but demonstrations of microscopically configurable trapping potentials have been lacking, with a notable exception being the production of multiple focused spots for the confinement of single atoms [8].

An alternative technique to manipulating the phase of the input beam is to instead utilise direct imaging. Though somewhat rarely encountered in optical tweezers, the technique known as *generalised phase contrast* uses the combination of a phase-based SLM and a phase-contrast filter to first create an amplitude pattern in an intermediate image plane, which is then directly reimaged in the optical tweezing plane [12]. The advantages of this technique are both speed and simplicity – the desired amplitude pattern can be directly written to the SLM without requiring the calculation of the appropriate hologram. This technique likewise avoids the generation of phase defects and speckle in the imaged pattern that can plague SLMs in the Fourier plane [5, 6], while being adaptable to the generation of large numbers of traps [12–14]. This comes at the cost of the ability to correct wavefront aberrations, but this disadvantage can be mitigated with a well-corrected optical system, as shown here. Another drawback is that the light efficiency is proportional to the fraction of illuminated trap area to maximum trap area.

A more recent addition to the toolkit for producing arbitrary optical potentials has been the digital micromirror device (DMD). Consisting of (up to) millions of individually addressable mirrors in a compact package, DMDs have the advantage of fast full-frame refresh rates on the order of 20 kHz, $\sim 20\times$ that of comparable liquid-crystal-based SLMs. DMDs can also be operated in a fixed fashion (DC) as they latch mirrors between reset pulses. Originally developed for digital light processing (DLP), these devices have seen increasing use in laboratory and industrial applications [15]. A DMD can be considered a dynamically configurable amplitude mask, which makes it highly suitable for direct imaging applications. These DMDs have been used to produce flattop beams for implementation into quantum gas experiments [16], incorporated into high-resolution systems for the purpose of single-site addressing in atomic quantum gas microscopes [9, 10], used to produce moving lattice potentials [17], and have recently been utilised to produce target-shaped traps [18]. DMDs may be used in either the Fourier plane [9, 11, 19–21] or directly imaged [10, 18]. Work by the Munich group has shown the usefulness of high-resolution direct imaging (600 nm FWHM) of a DMD in the creation of a two-dimensional disordered lattice for the exploration of many body localization transitions [22].

In this chapter, the basic understanding of DMDs is presented in terms of how they mechanically work and what their optical properties are. The ideas behind time-averaging and grayscaling are introduced followed by experimental results which demonstrate the utility of direct imaging of a DMD at high numerical apertures (NA) for optical trapping, which can be applied to trap a BEC. The optical system used for the experiments presented in this thesis has the major advantage of using commercially available optics and microscope objectives external to our glass vacuum chamber, and is corrected for the relatively thin 1.25 mm glass thickness is presented (see Chap. 3). This system demonstrates patterning of potentials with an upper-bound resolution of 630(10) nm FWHM at 532 nm illumination, within 5% of the diffraction limit for our 0.45 NA objective. The high-resolution potentials appear robust to tilts and misalignments of the objective and glass walls effects, in contrast to other cold atom experiments [23]. These patterns have an image extent of 130 μm $\times$ 207 μm in the atom plane, which allows nearly arbitrary sculpting of the optical potential and corresponding BEC. Subsequent imaging at the 780 nm resonant wavelength of ^{87}Rb achieves a submicron resolution of 960(80) nm FWHM, within 8% of the diffraction limit. Our DMD allows the storage of 13,889 frames and has a full-frame frame rate ranging from DC to 20 kHz, enabling diverse and dynamically configurable potentials. This rapid switching rate is used to introduce the use of DMDs for producing time-averaged potentials, which have been previously produced from rapidly scanning beams [24–26]. It is found that modulation frequencies above $\sim$3 kHz produce negligible heating of the atoms while allowing the production of six grayscale levels without halftoning. This technique can be combined with binary error-diffusion (halftoning) [16] to increase the number of grayscale levels available.

4.3 Operation

A DMD can be thought of as a massive array of small mirrors (called micromirrors) which are designed to rotate about their hinge axis (see Fig. 4.1). The binary mirror array is either in an square or diamond arrangement with a small gap between mirror elements for fabrication and mechanical reasons. These arrays of mirrors typically contain more than 1 million mirrors which are highly reflective to incident light. Each mirror is built on top of electrodes which apply the control field tilting the mirror by $\pm\theta_{\text{tilt}}$; these two binary states are called the 'on' and 'off' states. The mirrors are built on top of dual CMOS memory which is used to store the state of

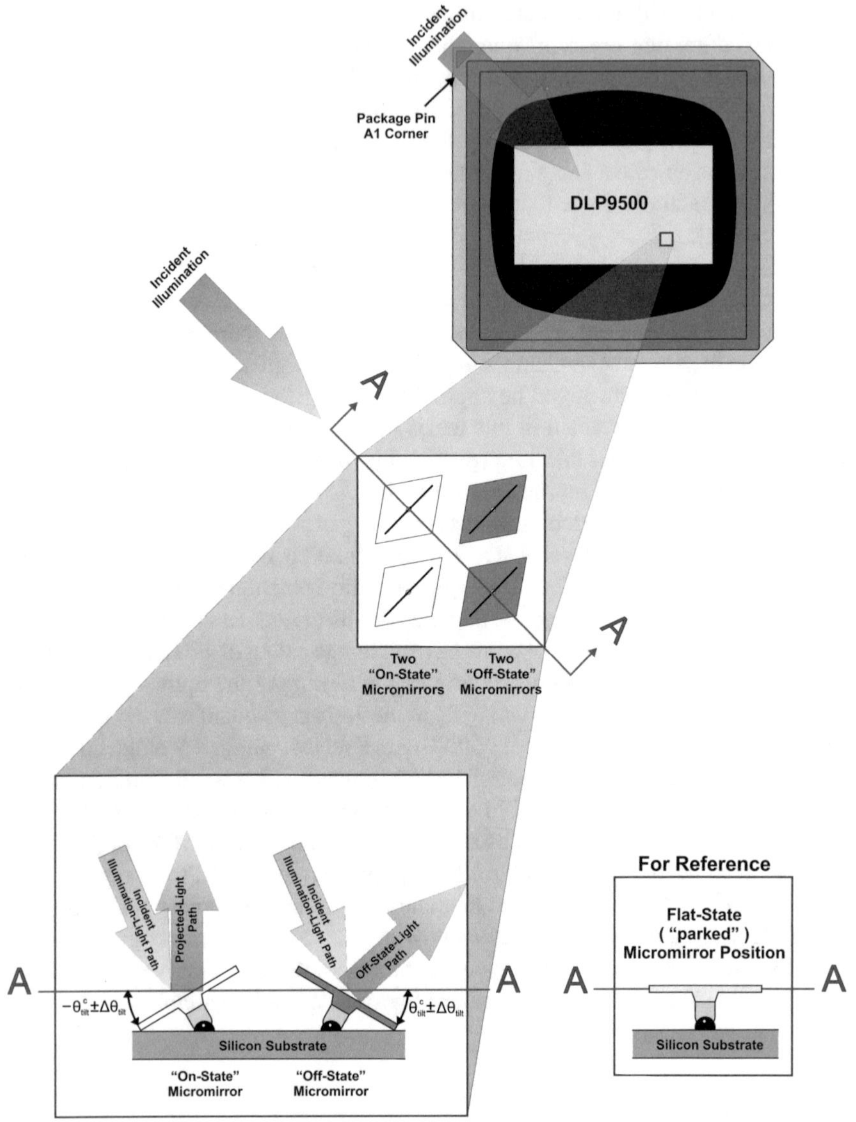

Fig. 4.1 Representation of the micromirror array and its mechanical components. Image originally from [28]

Table 4.1 Specifications of the DMD (DLP9500) [28] and controller (LUXBEAM 4600)

Micromirror array properties	
DMD array size	1920 × 1200
Micromirror pitch (d)	10.8 μm
Micromirror width[a] (a)	10.5 μm
Active region size	20.74 × 12.96 mm
Micromirror tilt angle	±12°
Micromirror tilt uncertainty ($\Delta\theta_{\text{tilt}}$)	±1°
AR coating visible light	400 to 700 nm
Micromirror array fill factor	94%
Micromirror array diffraction efficiency	87%
Micromirror reflectivity	89%
Window transmission	96%
Damage threshold	11 W/cm^2
Controller properties	
Onboard memory	4 Gb (13,888 1-bit frames)
Triggers	TTL, optical, 485 diff
Ports	Gbit ethernet
Pattern rate	20 kHz (1-bit)
	2.5 kHz (8-bit)

[a]The width of the mirrors is not specified by the manufacturer but we inferred it from the array fill factor and the micromirror pitch

the mirror switch to when a clock pulse is sent to the DMD, allowing for pre-loading of the image and a more simultaneous change of the image than if the mirrors changed state with CMOS memory upload. For a more complete introduction to DMDs see [27]. Note that the micromirrors rotate about their diagonal; this results in the best incident angle for illumination being along the diagonal of the mirror array. This can easily be accomplished by rotating the DMD array by 45°. This DMD design is influenced by the properties of 2D blazed grating diffraction, as explained below. The particular properties of the Luxbeam WUXGA 1200 × 1920 DMD used to project our potentials onto our atom plane are shown in Table 4.1.

Diffraction from the DMD The DMD mirror array can be thought of as a two-dimensional blazed diffraction grating with ±12° grating angle [29]. Blazed diffraction gratings, compared to standard gratings, have the advantage that the relative location of the diffraction orders and the centre of the reflected intensity peak envelope can be controlled independently. The position of the centre envelope can be shifted by $2\theta_g$, where θ_g is the angle of the grating, relative to the 0th order diffraction due to the blazing of the grating, allowing for the power to be localized in orders other than the 0th order. The blazed angle refers to the incident angle for which a given order diffracts from the grating at the same angle as the intensity peak leading to maximization of the power in that diffraction order. This allows for the grating

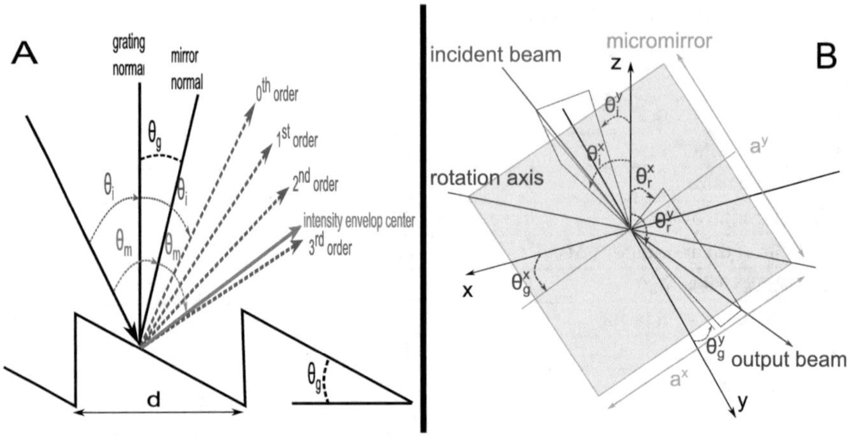

Fig. 4.2 Standard 1D blazed diffraction grating operation and 2D DMD grating operation and coordinates. **A** shows a 1D blazed diffraction grating with pitch d and grating angle θ_g. A light beam incident at θ_i with respect to the normal leading to the 0th order being reflected at $-\theta_i$ with other diffraction orders following the standard diffraction grating formula. The blazed grating angle decouples the intensity envelope from the 0th order of the diffraction grating since the light beam comes in with an angle θ_m with respect to the blazed surface which leads to a reflection at an angle of $-\theta_m$ with respect to the same normal or $(-\theta_i - 2\theta_g)$ with respect to the grating normal.**B** depicts the 2D diffraction grating that is the DMD with mirror size $a_y \times a_x$. Standard micromirrors have square dimensions $(a_y = a_x)$ and rotate about their diagonal. The diffraction condition can be treated the same as in the case of a 1D diffraction grating but needs to be applied along two directions. The mirror rotates about the rotation axis leading to an α_x (α_y) angular tilt about the y(x) axis. Due to the DMD mirrors being square and the tilt angle in both directions will be the same $\alpha_x = \alpha_y$. The incident beam projected onto the z-y(z-x) plane makes an angle $\theta_{y,i}(\theta_{x,i})$ with respect to the z-axis and the output beam projected onto the same plane makes an angle $\theta_{y,r}(\theta_{y,r})$ with respect to the same axis. Lastly, since we are working with a 2D diffraction grating the distance between the mirror centre in the $x(y)$-direction are denoted $\text{d}_x(\text{d}_y)$. Since the array is square the spacing in both direction is the same $\text{d}_x = \text{d}_y$

normal to be parallel to the optical axis when imaging the DMD, which avoids the Depth of Focus (DoF) issues that come from imaging a surface at an angle, while still keeping a high diffraction efficiency. This aspect is an advantage of DMDs over phase-based SLMs where the diffraction grating cannot be blazed.

For a 1D flat diffraction grating with slit spacing of d and incident wavelength of λ, the condition for constructive interference is given by

$$d\,[\sin(\theta_i) - \sin(\theta_n)] = n\lambda, \tag{4.1}$$

with n representing the diffraction order and θ_n representing the angles at which the orders is diffracted and θ_i is the incident light angle relative to the diffraction grating normal. We assume a positive angle convention where clockwise (anti-clockwise) angles relative to the grating normal are positive for the diffracted (incident) beam, see Fig. 4.2A. The 0th order is always reflected at $\theta_0 = \theta_i$ which also satisfies the reflection condition for a plane wave leading to the intensity peak and 0th order overlapping for a flat diffraction grating.

For a 1D blazed diffraction grating (see Fig. 4.2A) the diffraction orders can also be calculated using Eq. (4.1), but here the reflection happens with respect to the surface normal, which is the normal of the mirror, and not the grating normal. This results in the plane wave reflection angle not being the same as the 0th order diffraction angle. In this case, the peak happens at $\theta = \theta_i + 2\theta g$. This gives the ability to design gratings where most of the light does not end up in the 0th order but in some other order. The intensity profile for the blazed diffraction grating with N diffraction slits in the far field (Fraunhofer limit) gives a diffraction intensity profile of the form [30, 31]

$$I(\theta) = I_0 \left[\frac{\sin(\beta(\theta))}{\beta(\theta)}\right]^2 \left[\frac{\sin(N\alpha(\theta))}{N\sin(\alpha(\theta))}\right]^2, \tag{4.2a}$$

$$\beta(\theta) = \frac{ka}{2}\left[\sin(\theta_i + \theta_g) - \sin(\theta - \theta_g)\right], \tag{4.2b}$$

$$\alpha(\theta) = \frac{kd}{2}\left[\sin(\theta_i) - \sin(\theta)\right]. \tag{4.2c}$$

Equation (4.2a) gives the intensity in the far-field as a function of the output angle θ from the grating normal of a diffraction grating of N lines, d is the line spacing and a is the line width, k is the wave vector and is given by $2\pi/\lambda$ for plane wave illumination and θ_i is the angle of incidence relative to the surface normal.

In the case of a DMD, it consists of a blazed 2D diffraction grating with square slits in both directions which leads to an intensity in the far-field given by

$$I(\theta_x, \theta_y) = I_0 \left[\frac{\sin(\beta_x)}{\beta_x}\right]^2 \left[\frac{\sin(N_x\alpha_x)}{N_x\sin(\alpha_x)}\right]^2 \left[\frac{\sin(\beta_y)}{\beta_y}\right]^2 \left[\frac{\sin(N_y\alpha_y)}{N_y\sin(\alpha_y)}\right]^2, \tag{4.3a}$$

$$\beta_j(\theta_j) = \frac{ka_j}{2}\left[\sin(\theta_{j,i} + \theta_{j,g}) - \sin(\theta_j - \theta_{j,g})\right] \forall j \in \{x, y\}, \tag{4.3b}$$

$$\alpha_j(\theta_j) = \frac{kd_j}{2}\left[\sin(\theta_{j,i}) - \sin(\theta_j)\right] \forall j \in \{x, y\}. \tag{4.3c}$$

which is simply Eq. (4.2) applied to the light projected onto the y-z plane and projected onto the x-z plane with the same definition for the variables. Figure 4.2B depicts the 2D diffraction process. The diffraction orders are still given by $\alpha_j(\theta_{j,n}) = n\pi, n \in \mathbf{Z}$ and the reflection condition by $\theta_{j,r} = \theta_{j,i} + 2\theta_{j,g}$. One of the design features of the DMD is that the mirrors and array are square and the mirrors rotate about their diagonal which means that $a_x = a_y$, $d_x = d_y$, and $\theta_{x,g} = \theta_{y,g} \approx \theta_{\text{tilt}}/\sqrt{2}$, respectively, where θ_{tilt} is the tilt angle of the DMD about the rotation axis relative to the grating normal. This also means that for light incident perpendicular to the rotation axis $(\hat{x} + \hat{y})/\sqrt{2}$ when projected onto the x-y plane will have a reflected intensity envelope along the same x-y plane projected direction. This also means that the easiest way to use the DMD is by rotating it by 45°, since all of the diffraction orders $\theta_{x,n}(\theta_{y.n'})$ will also lie on the same axis for $n' = n$ making finding the blazing angle simple. Otherwise to get the maximum diffraction one requires a diffraction order where $n' \neq n$ so that the diffraction angle lines matches the reflection angle.

The way one wants to image a DMD is with the imaging light reflecting parallel to the grating normal so that the whole DMD is in the imaging plane of the optical system. This is done by setting the beam to come in perpendicular to the rotation axis at an angle that is twice the DMD tilt angle $\theta_{x,i} = \theta_{y,i} \approx \sqrt{2}|\theta_{\text{tilt}}|$. This means that for a mirror on the 'on' state $\theta_{\text{tilt}} = -12°$, and the intensity envelope is propagating along $\hat{z}$, and for an 'off' mirror state the intensity envelope is propagating at $4|\theta_{\text{tilt}}|$ from the z-axis along its original direction $(\hat{x} + \hat{y})/\sqrt{2}$. The blaze angle is defined as the incident angle at which the intensity envelope and an order of the diffraction grating overlap. The configuration above might not match the blazed angle condition, for the operating wavelength, in which case one needs to decide whether to maximizing the power of the or the diffraction limit of the optical system is the propriety. Since the incident angle can be optimized to center the maximum power diffraction order on the optical axis thereby minimizing off-axis aberrations, or it can be optimized to the blazed angle to maximize the power, but only for certain design wavelengths will the two angles match up.

It is interesting to look at what the expected diffraction efficiency for a given order is. To simplify the analysis, we assume that the grating is infinite (i.e. $N_x, N_y \to \infty$). This means that the only angles where light will be present are in the diffraction orders angles at $\theta_j = \arcsin\left[\sin(\theta_{j,i}) - n_j\lambda/d_j\right]$. We will also assume that the incident beam is perfectly aligned with $(\hat{x} + \hat{y})/\sqrt{2}$ in the x-y plane such that $\theta_{x,i} = \theta_{y,i} \approx \theta_i/\sqrt{2}$. Under these assumptions, the diffraction efficiency for each order is given by

$$\eta(m_x, m_y, \theta_i, \theta_{\text{tilt}}) = C(\theta_i)\,\text{sinc}^2\left[\beta_{x,m_x}\right]\text{sinc}^2\left[\beta_{y,m_y}\right], \tag{4.4a}$$

$$\beta_{j,m_j} = \frac{ka}{2}\left[\sin\left(\frac{\theta_i + \theta_{\text{tilt}}}{\sqrt{2}}\right) - \sin\left(\arcsin\left[\sin\left(\frac{\theta_i}{\sqrt{2}}\right) - \frac{m_j\lambda}{d}\right] - \frac{\theta_{\text{tilt}}}{\sqrt{2}}\right)\right], \tag{4.4b}$$

$$C(\theta_i) = \left[\sum_{m_x=n_-}^{n_+}\sum_{m_y=n_-}^{n_+}\text{sinc}^2\left[\beta_{x,m_x}\right]\text{sinc}^2\left[\beta_{y,m_y}\right]\right]^{-1}, \tag{4.4c}$$

$$n_- = \left\lceil\left(\sin\left(\frac{\theta_i}{\sqrt{2}}\right) - 1\right)\frac{d}{\lambda}\right\rceil, \tag{4.4d}$$

$$n_+ = \left\lfloor\left(\sin\left(\frac{\theta_i}{\sqrt{2}}\right) + 1\right)\frac{d}{\lambda}\right\rfloor, \tag{4.4e}$$

where $j \in \{x, y\}$. $C(\theta_i)$ is the normalization factor since the incident power must be conserved and the diffraction efficiency should sum to one over all the possible diffraction orders, bounded by $n_\pm$. In the case where the light is reflected normal to the surface, and assuming that most of the light is contained in the $m_x = m_y = m$ orders, the diffraction efficiency for that order is given by

$$\eta(m) = C\,\mathrm{sinc}^4\left[\frac{a\pi}{\sqrt{2}}\left(\frac{2\theta_{\mathrm{tilt}}}{\lambda} - \frac{m\sqrt{2}}{d}\right)\right], \tag{4.5}$$

where small angle approximation is used to simplify the equation. The same equation was derived in ref. [32] using Fourier optics.

Using the Table 4.1 and Eq. (4.4a), for the 532 nm light, we calculate a diffraction efficiency of 95.7% and using Eq. (4.5) we get ∼100%. The measured experimental efficiency is about ∼40% which even when taking other factors into consideration are quite far from one another. One of the factors that we did not take into consideration is the uncertainty of the tilt angles. The full treatment of this involves keeping track of the phase difference across the DMD surface during diffraction which could be quite complex. As a first-order approximation, we will simply treat the DMD as made up of a superposition of diffraction gratings with a probability distribution given by

$$\rho(\theta_{\mathrm{tilt}})d\theta_{\mathrm{tilt}} = \sqrt{\frac{2}{\pi\,\Delta\theta_{\mathrm{tilt}}^2}}\,\exp\left[-2\left(\frac{\theta_{\mathrm{tilt}} - \theta_{c,\mathrm{tilt}}}{\Delta\theta_{\mathrm{tilt}}}\right)^2\right]d\theta_{\mathrm{tilt}}, \tag{4.6}$$

which is a Gaussian distribution centred about $\theta_{c,\mathrm{tilt}}$ and with 95% confidence interval given by the specified DMD micromirror tilt uncertainty ($\Delta\theta_{\mathrm{tilt}}$). To obtain the diffraction efficiency of a given order we calculate the diffraction efficiency to be

$$\eta(m_x, m_y, \theta_i) = \int_{-\pi/2}^{\pi/2} \eta(m_x, m_y, \theta_i, \theta_{\mathrm{tilt}})\rho(\theta_{\mathrm{tilt}})d\theta_{\mathrm{tilt}}, \tag{4.7}$$

where $\eta(m_x, m_y, \theta_i, \theta_{\mathrm{tilt}})$ is given by Eq. (4.4a). For our DMD the diffraction efficiency then becomes ∼74% in the $n_x = n_y = 6$ order. Now accounting for the mirror reflectivity, array fill factor and window transmission (see Table 4.1), we get a theoretical diffraction efficiency of 57%. If the specified $\Delta\theta_{\mathrm{tilt}}$ is a 60% confidence interval measurement instead of 95% as assumed (not specified in the datasheet) then expected diffraction efficiency of the array is calculated to be 50% in the same order. To first-order, the uncertainty in the mirror tilt angle is very detrimental for the diffraction efficiency of the array.

Polarisation effects The polarisation of the input and output beam from the DMD were measured and found to be maintained contrary to what was reported in [33] where they found an elliptical polarisation output for a linear polarisation input. We do however notice a change in the total diffraction efficiency of the DMD by going from vertically to horizontally polarised light of about ∼10%. To avoid drift of the input fibre polarisation causing shot-to-shot issues, we pass the output of the fibre through a polarising beam splitter cube before stabilizing the 532 nm power.

4.4 DMD Atomic Patterns and Half-Toning

In this section, we first introduce the concept of halftoning and how it is related to the diffraction limit of the imaging system. We then measure the point spread function (PSF) of the atom imaging system and DMD projection system using light. Subsequently, we show images of actual atoms loaded into DMD created light patterns and use the Siemens Star resolution target to measure the modulation transfer function to put an upper limit on the pattern resolution achieved at our atoms, and use the same method to more accurately measure the resolution of the DMD projection.

4.4.1 Projecting with Light

As the DMD mirrors have only two states — 'on' and 'off' — the resulting image is binary. However, given the limited spatial resolution of a typical optical system, the technique known as error-diffusion or halftoning can be used to produce intensity gradients [16, 18, 34]. We determine the resolution of our system by first turning on a single mirror of the DMD and projecting it on to the atom plane. The mirror pitch is 10.8 μm, so the minification factor of 100 results in an projected mirror width of 108 nm, below the theoretical resolution limit of the top microscope system (605 nm FWHM at 532 nm illumination). As the single mirror is not resolvable, its image at the atom plane approximates the point-spread function (PSF) of the DMD imaging system. We reimage this focused spot on the camera with the bottom microscope system. Accounting for the magnification factor of the reimaging system results in a 650(50) nm FWHM peak in the atom plane, as shown in Fig. 4.3A. This value is an upper limit, as it convolves any aberrations of the reimaging system into the estimate of the spot size at the atom plane. Back-propagating this resolution element to the DMD location gives a 65 μm spot which spans a $\sim 6 \times 6$ block of mirrors; multiple mirrors will thus contribute to the resolution spot in the atom plane [16]. We therefore can use error-diffusion to control the light intensity at the atom plane, as shown in Fig. 4.4B.

The imaging resolution limit for ^{87}Rb atoms can also be measured by again turning on a single mirror of the DMD, but illuminating it at 780 nm. We can estimate the PSF at this wavelength, as shown in Fig. 4.3B. This results in a 960(80) nm FWHM peak, consistent with the measurement at 532 nm after accounting for the increased wavelength. Since the imaging of the atoms will be carried out using the same system, we expect the atomic patterns to be imaging resolution limited and not projected pattern limited.

The resulting high-resolution DMD patterns produce a repulsive potential for the atoms trapped in the sheet trap due to the blue-detuned projection wavelength (532 nm) used in projecting the DMD onto the atom plane. The light power incident on the atoms can be varied from no light all the way to 7μ, where μ is the chemical potential of $\sim 50\,nK \times k_B$. However, we observed a decrease in the BEC lifetime

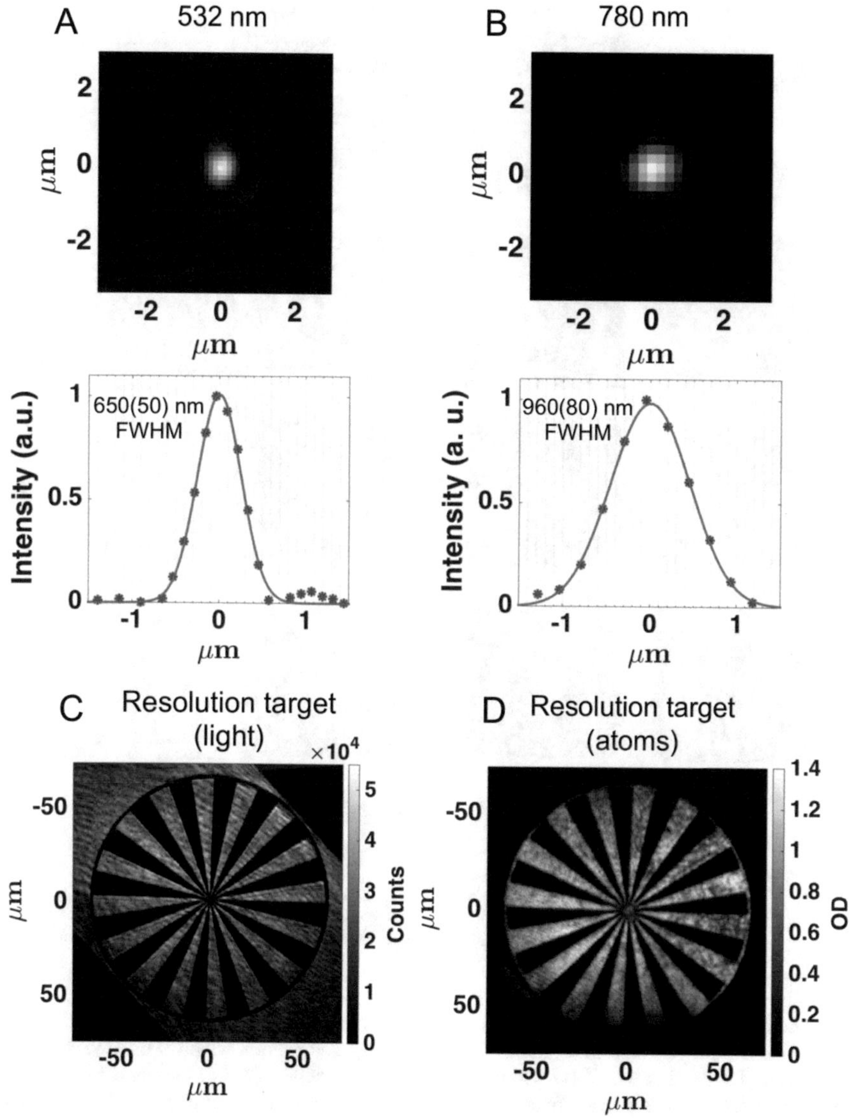

Fig. 4.3 **A** A single DMD mirror imaged on to the camera with 105× magnification and 532 nm illumination, resulting in 650(50) nm FWHM ($\omega_0 \sim 550$ nm). The FWHM was determined through 100 1D fits of the image at varying angles through 180 degrees; a single 1D fit is shown here for illustrative purposes. **B** A single DMD mirror imaged with 52.6× magnification and 780 nm illumination, the imaging wavelength for ^{87}Rb, resulting in 960(80) nm FWHM ($\omega_0 \sim 814$ nm). **C** Siemens star resolution target, as projected and reimaged with 532 nm light. **D** Siemens star imprinted on to the atomic density and averaged over 10 runs of the experiment. Quantitative analysis of the resolution targets is presented in Fig. 4.5

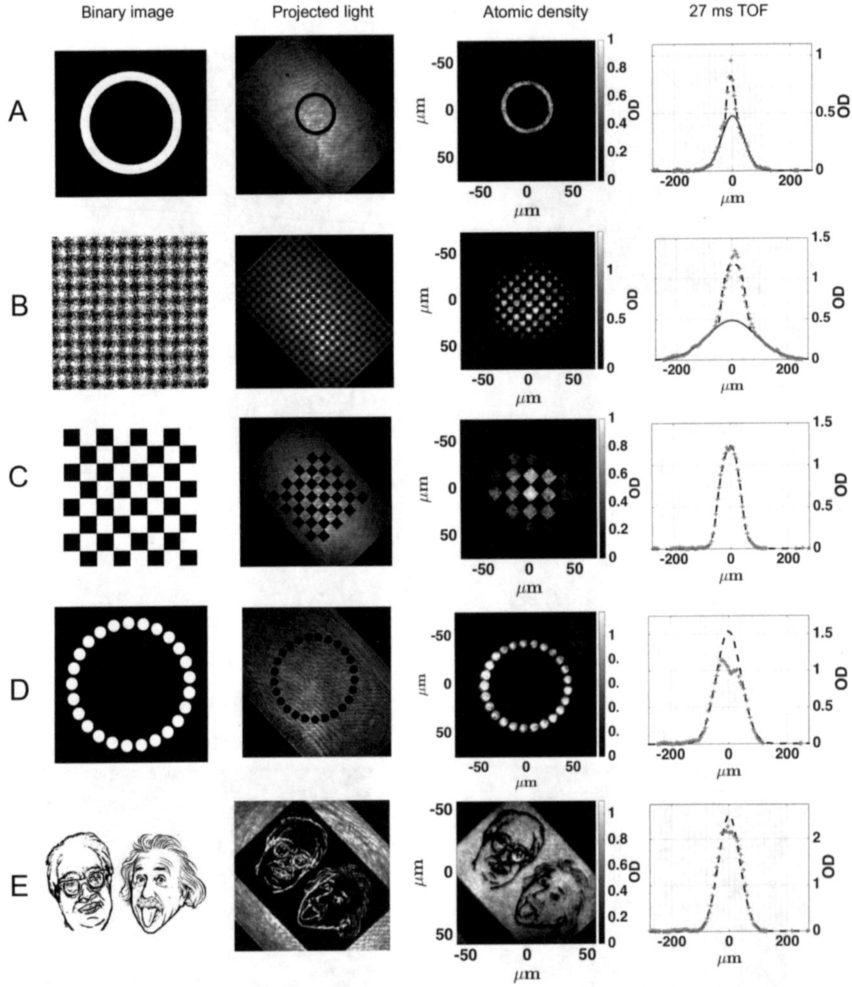

Fig. 4.4 DMD patterned optical traps and resulting resonant absorption images of atom distributions. Atoms are repelled from bright regions of the projected pattern, which is the inverse of the binary image applied to the DMD. We image the atoms immediately after turning off the optical trapping potentials, with a magnification of 52.6×. Bright areas represent regions of high atomic density and optical depth (OD). **A** Ring trap potential from a single experimental realization. Time of flight analysis (TOF) gives $N = 1.3 \times 10^5$ atoms with matter-wave interference leading to the appearance of a central peak [40]. **B** Lattice pattern produced by applying a Floyd-Steinberg error diffusion algorithm [34] to an 8-bit image of a sinusoidal lattice with 10 μm period, shown with a single realization of a BEC; $N = 3.7 \times 10^5$ and the BEC fraction is 34%. This image was produced by leaving the magnetic trap on, resulting in $\omega_r = 2\pi \times 20$ Hz harmonic radial confinement. **C** Checkerboard pattern applied to the atoms, imaged with a single shot. Additional evaporation after transfer to the all-optical trap results in a nearly pure BEC in TOF with $N = 2.4 \times 10^5$ atoms. **D** Ring lattice of 25 sites, with a ring radius of 43.2 μm, and site radius of 4.32 μm, with $N = 3.16 \times 10^5$ atoms. ODs above ~2.5 are below the signal to noise threshold of the horizontal-imaging, leading to slight undercounting of the atoms in this case. **E** Artistic impressions of Bose and Einstein (Einstein image from www.muraldecal.com, used with permission.) applied to a nearly pure Bose-Einstein condensate of $N = 5.2 \times 10^5$ atoms, averaged over 5 experimental runs

with increasing intensity, with measured time constants of (18.2, 10.9, 9.1, 8.2) s for an 80 μm × 50 μm rectangular DMD box potential with peaks of (1.2μ, 2.4μ, 4.8μ, 7.2μ), respectively. Similar intensity-dependent lifetimes were observed for different trap geometries. This dependence is partially explained in [35], where it is pointed out that to prevent latching of the DMD micromirrors on the surface when they are on for too long, the manufacturer implements a relaxation cycle every MCP clock cycle. This cycle causes noise on the order of 50 μs which might cause degradation of the lifetime, and is more significant at higher power. The effect was more pronounced in their case as they were using is ^{6}Li which has a lower mass than ^{87}Rb. They present a way to work around this problem by interrupting the clock pulses to hold the mirrors latched until the next frame needs to be displayed [35].

Several high-resolution atom-trap configurations are shown in Fig. 4.4, emphasizing the wide range of possible patterns. These are imaged immediately after turning off the optical trap. Alternatively, by imaging from the side after 27 ms time of flight (TOF) expansion, we can determine atom number and BEC fraction for the different distributions. We note that we have used the same evaporation ramp for each of these configurations — we expect that a near pure BEC can be achieved for each with optimization. We generally observe lower condensate fractions for smaller enclosed areas for similar optical evaporation profiles.

We briefly describe examples of the trapping potentials we have generated. Figure 4.4A demonstrates a ring trap, of interest for atom interferometry [36] and studies of phase slips and persistent currents [37, 38]. Figure 4.4B shows a 10 μm-period optical lattice. The binary DMD pattern was generated by applying a Floyd-Steinberg error diffusion algorithm [34] to an 8-bit grayscale image of a sinusoidal lattice. As we have not included a boundary to the lattice in the DMD pattern, we leave the magnetic field on and retain a $2\pi \times 20$ Hz radial trapping frequency, producing a more symmetrically filled lattice. Figure 4.4C demonstrates a checkerboard pattern applied to the BEC, and the evaporation ramp, in this case, results in a BEC with a negligible thermal component. This pattern emphasizes the sharp features in the atomic distribution resulting from the highly resolved features of the DMD pattern. Ring lattices have generated wide interest [39], and Fig. 4.4D demonstrates a ring lattice of 25 sites. Finally, in Fig. 4.4E we project artistic impressions of Bose and Einstein into a nearly pure BEC, resulting in a "Bose-Einstein" Bose-Einstein condensate and demonstrating our ability to create arbitrary potentials. This image is an average over 5 experimental runs (30 s experimental cycle) and is similar to Fig. 4.3D which uses a 10-run average. These images demonstrate the repeatability of the high-resolution patterned BECs. The integrated optical density (OD) TOF cross sections in Fig. 4.4D, E deviate from Gaussian fit at their centre. In the case of Fig. 4.4D, this is due to an uneven density distribution of atoms across the ring lattice and insufficient spatial overlap, or incoherent overlap of the lattice sites after only 27 ms TOF. In the case of Fig. 4.4E, this deviation comes due to the high atomic densities leading to ODs above $\sim$2.5 which are below the signal to noise threshold of the horizontal-imaging. In each of these cases, the limited expansion of the cloud, when compared with the thermal component of Fig. 4.4B, along with the lack of a bimodal distribution, indicate a relatively pure BEC state.

4.4.2 Modulation Transfer Function, MTF, from Light and Atomic Distributions

For a more quantitative characterisation of the optical systems [41], we produced a binary Siemens Star resolution target on the DMD. We then imaged both the projected light pattern, and the BECs loaded into the pattern, as shown in Fig. 4.3C, D. These patterns can then be analyzed to determine the modulation transfer function (MTF) of the total optical system. This is accomplished using the protocol illustrated in Fig. 4.5. We first determine circular paths around the Siemens star, with frequency spacings $\Delta = (0.047, 0.034)$ line-pairs (lp)/μm, for the (light, atom) Siemens star. A contrast value is found for each of the 16 adjacent bright and dark spoke pairs, and the average contrast is calculated for each circular path. As the MTF is the Fourier transform of the PSF, which we estimate with a Gaussian fit, we extract the FWHM from a Gaussian fit of the measured MTF. These quantities are then related by $\text{FWHM}_{\text{PSF}} = 4\ln(2)(\pi \text{FWHM}_{\text{MTF}})^{-1}$, allowing us to compare the PSF FWHM extracted from the MTF with that of the single mirror image. For the Siemens star with 532 nm light, we find a MTF FWHM of 1.39(0.02) lp/μm, corresponding to a PSF FWHM of 630(10) nm, in agreement with the 650(50) nm measurement of the single mirror image. When imaging a BEC in the Siemens star pattern, we find a MTF FWHM of 0.71(0.02) lp/μm, corresponding to a PSF FWHM of 1250(20) nm, larger than the 960(80) nm single mirror image at 780 nm illumination. We believed this ~30% increase is consistent with atom diffusion due to photon recoil during the 10 μs repump and resonant imaging pulse [42]. To verify this we repeated the same procedure but this time using Faraday imaging (Sect. 2.5.3) which interacts with the birefringence of the cloud instead of through scattering and measured reduced PSF FWHM of 1080(15) nm which confirms that part of the broadening was due to the recoil due to the imaging method itself.

4.4.3 Time-Averaged Potentials with DMDs

The DMD is capable of switching mirrors from DC to the specified maximum frequency of 20 kHz, which we have verified through photodiode measurement. This wide modulation range, along with the ability to store 1,700 frames on the device, enables a high level of dynamic control. At low modulation frequencies, this allows adiabatic deformation of the DMD potentials. At the other extreme, high-frequency switching can be utilised for quickly quenching the potential geometry. Furthermore, the painted-potentials technique is possible, where an average dipole force is produced through rapid modulation of the optical field [24, 25]. This suggests pulse-width-modulation (PWM) as an alternative technique to error diffusion for producing grayscale levels, analogous to the techniques used in DLP.

To explore the suitability of this system for producing time averaged potentials, we produced the DMD pattern shown in Fig. 4.6, consisting of an array of six 8 μm

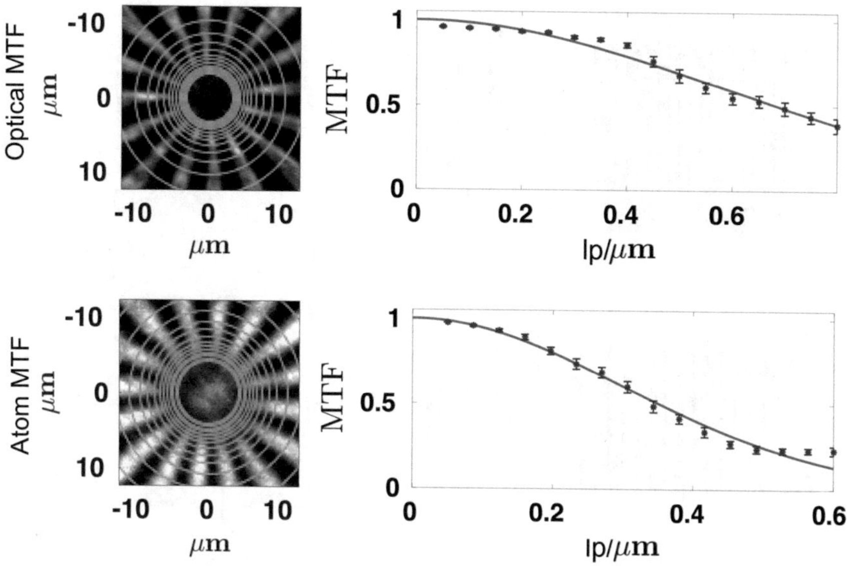

Fig. 4.5 Modulation transfer function (MTF) analysis. Left: zoomed versions of the Siemens star images in Fig. 4.3, with the top row representing the optical pattern at 532 nm illumination, and the bottom an average atomic density in the pattern, imaged with resonant 780 nm light. The circles indicate radii of equally separated spatial frequencies used to generate the corresponding MTF plots by calculating contrast along the circular path. Right: the FWHM of the optical pattern is 1.39(0.02) lp/μm, corresponding to a PSF FWHM of 630(10) nm. The atomic density MTF FWHM is 0.71(0.02) lp/μm, corresponding to a PSF FWHM of 1250(20) nm

diameter barriers contained in a 50 μm × 80 μm rectangle. PWM, with the DMD running at an envelope frequency of $f_e = 2.75$ kHz, was utilised, and the leading edge of each pulse was fixed. The hexagonal array within the rectangle was modulated with varying duty cycle over six levels by subdividing the carrier pulse into six modulation divisions, with the maximum duty cycle corresponding to all divisions turned on, and the minimum corresponding to one division turned on. The carrier frequency thus used was $f_c = 16.5$ kHz. We illuminated the DMD with light corresponding to a trap depth of 1.2μ. The corresponding relative atomic densities are shown in Fig. 4.6C.

To investigate the limits of this technique, we measured the effect on temperature and BEC fraction due to modulation of a single barrier (labeled "3" in Fig. 4.6) which had a 50% duty cycle. After first forming the BEC in the rectangle in the absence of any internal barrier, we performed a modulation for 500 cycles at fixed frequencies ranging from 100 Hz to 16 kHz. We observed a decreasing effect as the frequencies are increased due to net energy gain per cycle decreasing since the barrier oscillates on a smaller time scale than atom diffusion time scales; above ~3 kHz the effect is negligible.

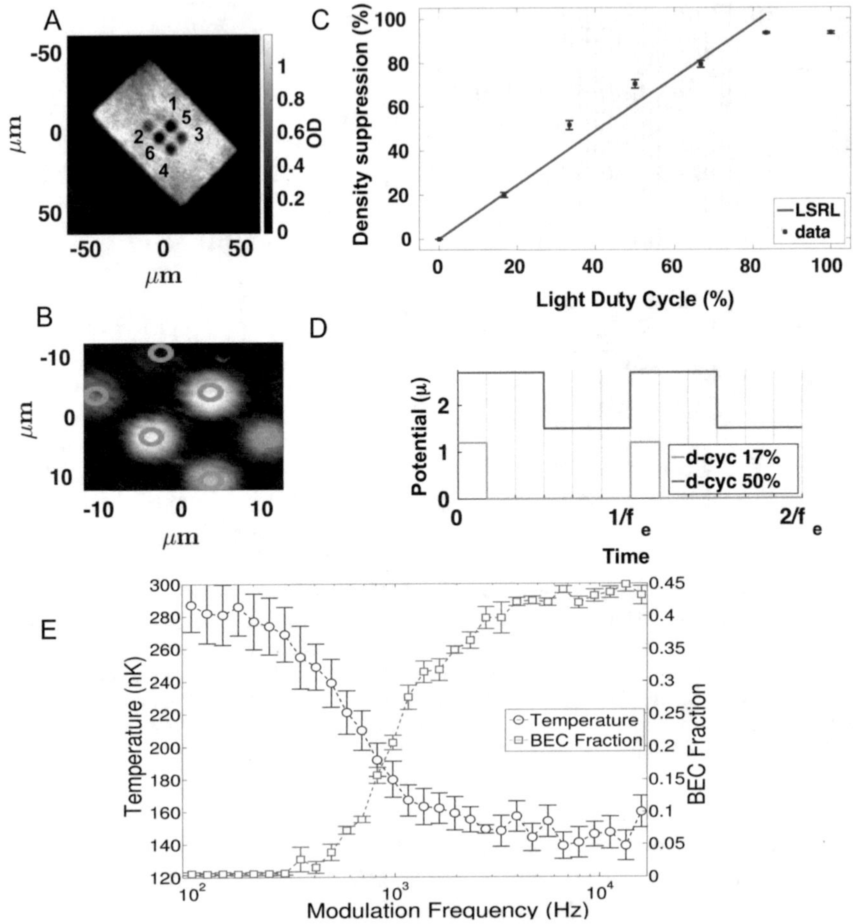

Fig. 4.6 **A** Grayscale time-averaged pattern applied to the atoms, averaged over 10 sequential images. Each of the numbered patterns corresponds to the fractional duty cycle of 2.75 kHz, with 6 being the maximum (see text), corresponding to a potential depth of 1.2 μ. **B** Images used for analysis of the gray levels. The grayscale image was subtracted from the average atomic background with no barriers present. The mean density and standard deviation were calculated over the circles as indicated. **C** Gray levels achieved through this process. A least squares regression line (LSRL) is indicated. As the optical potential of point 6 exceeds the condensate chemical potential it was excluded from the fit. Image background and untrapped atoms result in an apparent maximum density suppression of 96%. **D** Two example pulse width modulations, corresponding to patterns 1, the minimum pulse, and 3, a 50% duty cycle, offset vertically for clarity. The envelope frequency $f_e = 2.75$ kHz and carrier frequency $f_c = 16.5$ kHz divisions are indicated. **E** Turning on and off barrier 3 only, with varying frequency, for a total modulation time of 500 cycles provides an estimate of the heating rate from mirror switching. Rates above ~3 kHz appear to have a negligible heating effect over this modulation period

4.5 Feed Forward

When projecting complicated halftone potentials, we use a modified Floyd-Steinberg algorithm which accounts for the intensity of the incident light on the DMD to generate our binary patterns. However, due to imperfections in the incident light and aberrations/dust in the DMD projection system, the DMD projected potential we obtain is not exactly the expected potential. These defects in the trapping potential can be corrected by either feeding back on the light signal or on the atomic signal, which is representative of the confining potential in the Thomas-Fermi limit, see Sect. 2.2.4. Both of these methods have advantages. In the cases of using the light, the feedback happens comparatively quickly since one is only limited by the speed at which the camera takes images and the speed at which they are uploaded to the computer. When using the BEC density, the limiting factor is how quickly the apparatus can generate data, typically giving an image every 30 s. When feeding back using the light signal, one of the drawbacks is that one must differentiate between light pattern imperfections that are due to an aberrations in the imaging system and aberrations in the projection system. One only wants to feedback for imperfections in the potential due to the projection system else imperfections are added to the potential in the atom plane. The same drawback exists when feeding back using the atomic density which is mitigate due to measuring the atoms reaction to the potential instead of the potential itself. Imaging the BEC has the advantage of only needing one optical system instead of needing two imaging systems, one for the potential light and one for the atoms.

In this section, the density-based feedback algorithm used to correct the patterns is introduced. The algorithm primarily consists of three steps. The first step is to generate a mapping between the camera coordinates and the DMD pixels. To do this we use image recognition to create a best-mapping between the camera and DMD, which is calculated for every image, avoiding drift problems associated with static methods. The second step is to generate the error map for the DMD; it is a measure of the distance the density at each pixel is from the target density. Finally, we modify the DMD pattern in a prescribed fashion so that the error in the next iteration is lower than the previous one. The process can then be repeated over and over until the desired pattern is achieved.

4.5.1 Image Recognition

The key to good image recognition is a good initial guess, true for any numerical based optimization. In this case, the initial guess is the DMD pattern (which is generated and moved around) to try and fit to the target image (which is not modified). The way we generate our first guess is simply using a best-known transform between the DMD pixels and camera pixels. For our experiment, we know that the camera and DMD are rotated $\theta \sim 45^\circ$ from one another, that the pattern is inverted through the imaging system, and that the DMD and camera are centred on the atom cloud. Lastly, the DMD has a magnification of M_{DMD}, at the atom plane, and a corresponding pixel size of p_{DMD}. Similarly, we know that the camera has pixel size p_{CAM} and magnification M_{CAM} from the atom plane to the CCD. The initial transform between the two is thus given by

$$\mathbf{A} = \begin{bmatrix} \frac{p_{DMD} M_{CAM}}{p_{CAM} M_{DMD}} & 0 & 0 \\ 0 & \frac{p_{DMD} M_{CAM}}{p_{CAM} M_{DMD}} & 0 \\ 0 & 0 & 1 \end{bmatrix} \begin{bmatrix} \cos\theta & -\sin\theta & 0 \\ \sin\theta & \cos\theta & 0 \\ 0 & 0 & 1 \end{bmatrix} \begin{bmatrix} -1 & 0 & 0 \\ 0 & 1 & 0 \\ 0 & 0 & 1 \end{bmatrix} \tag{4.8}$$

where the first matrix maps DMD pixels to camera pixels in term of size, the second matrix undoes the rotation and the third mirrors the system about the x-axis. It is worth noting that we assume the origin of the image is the centre of the image, while most image recognition software assumes the origin is the top-left corner. In this case to obtain a rotation one needs to translate the centre of the image to the origin and perform the rotation, before translating the image back to its original coordinate frame:

$$\mathbf{A}_{\text{rot}} = \begin{bmatrix} 1 & 0 & 0 \\ 0 & 1 & 0 \\ -N_x/2 & -N_y/2 & 1 \end{bmatrix} \begin{bmatrix} \cos\theta & -\sin\theta & 0 \\ \sin\theta & \cos\theta & 0 \\ 0 & 0 & 1 \end{bmatrix} \begin{bmatrix} 1 & 0 & 0 \\ 0 & 1 & 0 \\ N_x/2 & N_y/2 & 1 \end{bmatrix} \tag{4.9}$$

where $N_x(N_y)$ is the number of pixels along the $x(y)$-direction of the image. Here, the first and third matrices accomplish the translations back and forth. Note that the last row of the rotation matrix should always be 0, 0, 1 since the image adaptation is in 2D.

Usually, the target image is assumed to be a uniform density profile, given by the inverse of the DMD binary pattern as in Fig. 4.7A, B. In the case of an halftoned or time-averaged DMD potential, the Thomas-Fermi of the time-averaged potential is used as the density target as shown in Fig. 4.8A, B.

Once the best guess density from the DMD pattern is found it can be converted to an expected density pattern in the camera frame of reference using Eq. (4.8) and the imaged density and camera density can be midway normalized (see Sect. 4.5.1) to one another. This gives a guess density profile in the camera frame of reference as depicted in Figs. 4.7C and 4.8C. This guess can be compared to the imaged density (Figs. 4.7D and 4.8D) with the overlap shown in Figs. 4.7E and 4.8E. An iterative process is then used to minimize the least mean square (LMS) error between the two images, with each step consisting of optimizing the overlap between the target (measured density) and the guess. The guess is then renormalised to the target density profile, using midway normalization (described below), with the new overlap. The process is repeated until the LMS is converged to within a certain tolerance. At first, we only allow for translation of the guess during the optimization. Once the translation has been optimized, we allow for translation and rotation of the image using the results of the translation as the starting guess for our optimization. Finally,

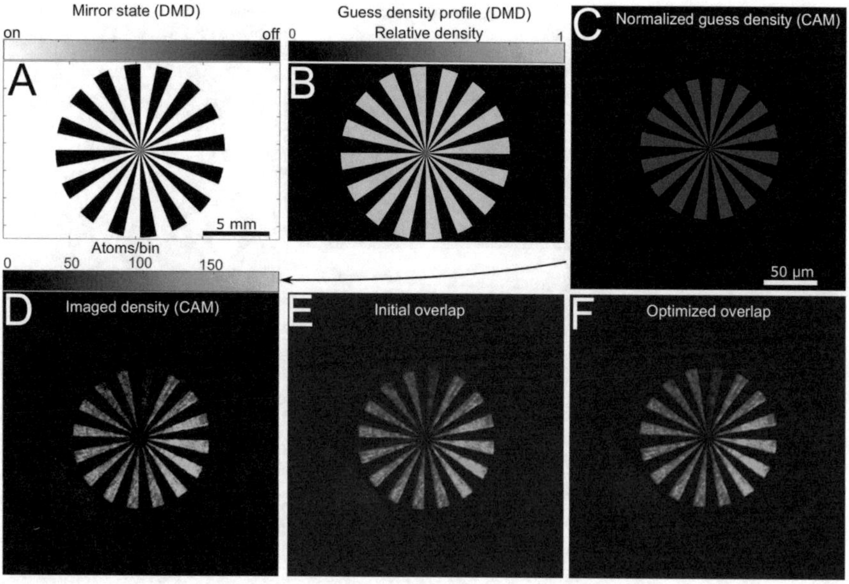

Fig. 4.7 Binary pattern image recognition. **A** shows the uploaded DMD pattern with dark(white) representing a region with the mirrors in the 'off'('on') state and white representing mirror in the 'on' state. **B** is the guess density profile due to the projection of **A** on the atom plane. **C** is the transformed guess profile using Eq. (4.8) and midway normalized to **D**. **D** measured density using Faraday imaging. **E** is the initial overlap between the guess and the actual pattern and **F** is the overlap after optimization where green represents the measured signal and purple is the guess density

we further add scaling as a degree of freedom to the problem, once again using results of the last optimization as the starting guess. Once the optimization is complete with specified tolerance, we have 'recognized' the image and have effectively found the best transform between the DMD frame and the camera frame, which is the first step in creating a successful feedback loop. The results of this optimization are shown in Figs. 4.7F and 4.8F through the overlap of the guess and measured density image. The minimization of the LMS error is done using a steepest descent algorithm, for which MATLAB offers an easy to use function (imregtform) in its image processing toolbox, which will perform the LMS optimization and can also perform other types of image overlap optimization.

Note that another way to perform the same operation is to use a reference pattern of atoms that will be the same from image to image, such as is done with positioning patterns used in certain lithography applications [43]. This pattern can be generated by the DMD and provides an easy way to match the two images, the main drawback is the reduction in the usable DMD region over which the potentials can be drawn.

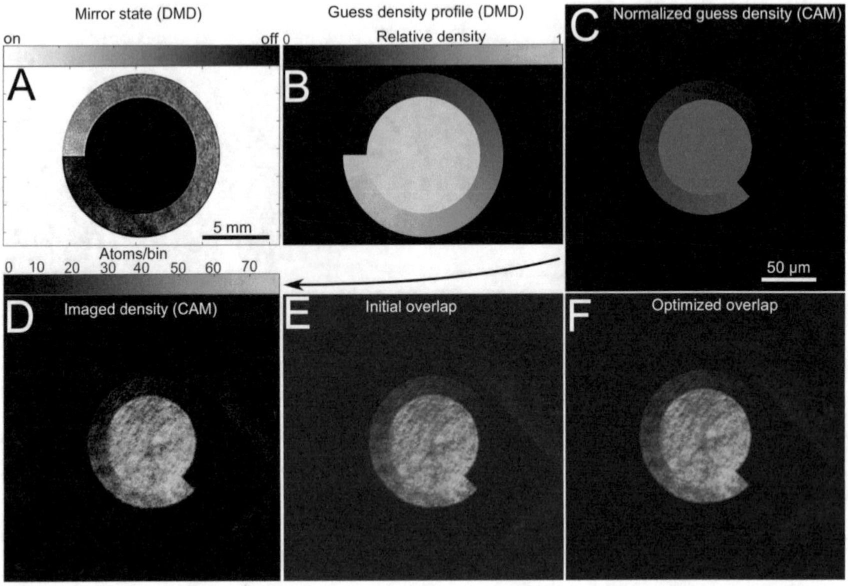

Fig. 4.8 Halftoned pattern image recognition. **A** shows the uploaded DMD pattern with dark(white) representing a region with the mirrors in the 'off'('on') state and white representing mirror in the 'on' state. **B** is the density pattern we are trying to recreate through halftoning. **C** is the transformed guess profile using Eq. (4.8) and midway normalized to **D**. **D** measured density using Faraday imaging. **E** is the initial overlap between the guess and the actual pattern and **F** is the overlap after optimization

Modified Midway Image Equalization

It is a lot simpler to work using relative densities and signal when doing image recognition since the absolute numbers may change from shot-to-shot experimentally. To get around this issue it is possible to use midway equalization. This process of normalizing or equalizing one image to another is used in a variety of fields and applications where images from different sensors/cameras need to be compared or analysed. It is also used when comparing multiple images, from the same sensor, that are separated in time over the course of which the system parameters are expected to have changed. The basic idea of most equalization code is to change certain parameters of interest in the image while preserving as much as possible the dynamics of the image. Some of the most common are range normalization where two images are normalized such that their min and max are the same, or mean normalization where the mean of the images are set to be the same. More advanced algorithms include histogram-wrapping and midway equalization whose goals are to normalize the histogram and cumulative histograms, respectively. For our image recognition, we use midway equalization combined with range normalization which requires no spatial information of the image, can be used with images of different size, and can be used to normalize the images using only a subset of the image.

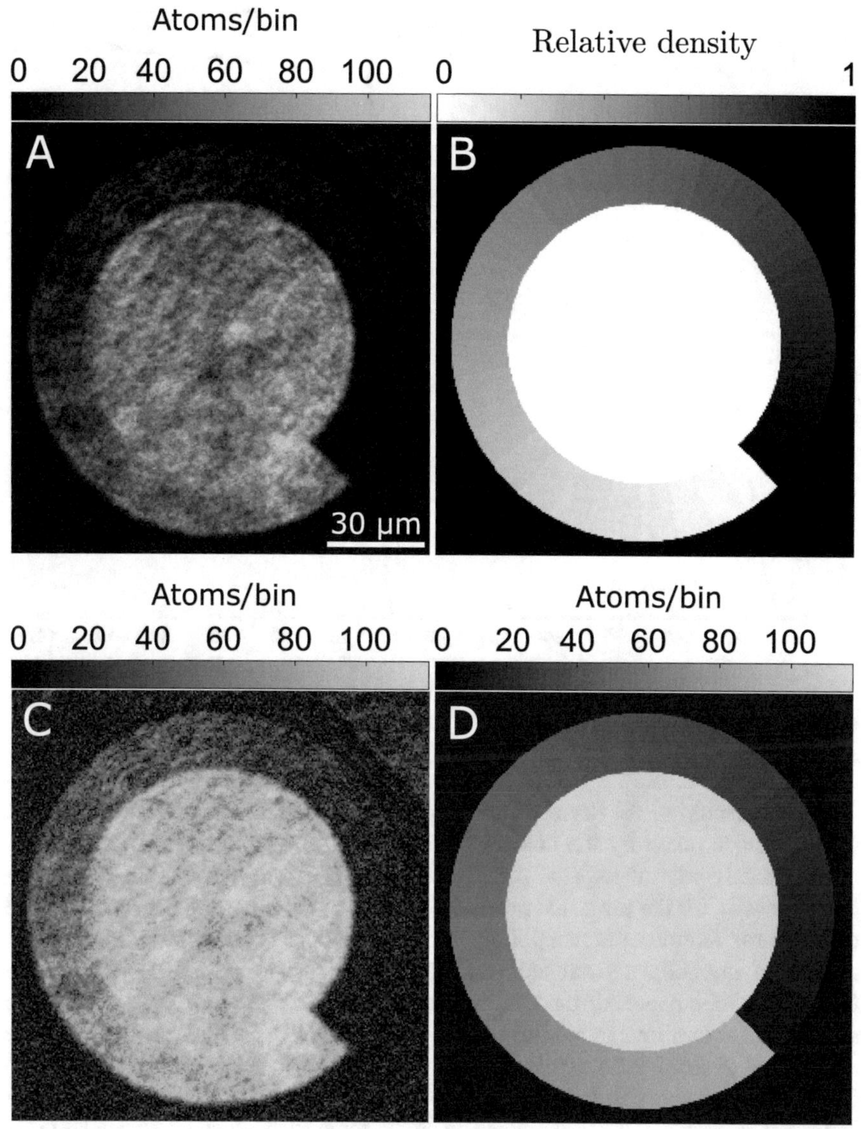

Fig. 4.9 Midway image equalization. **A** shows the camera density image and **B** the density profile guess specified in terms of relative density. **C**, **D** are the midway equalized images corresponding to **A**, **B** respectively with the data range set by **A**. **E**, **F** are the cumulative fractional histogram of the image and guess measured before, and after midway normalization which was performed using algorithm presented in [46]. **G**, **H** are the same as **E**, **F** but using our modified midway normalization code

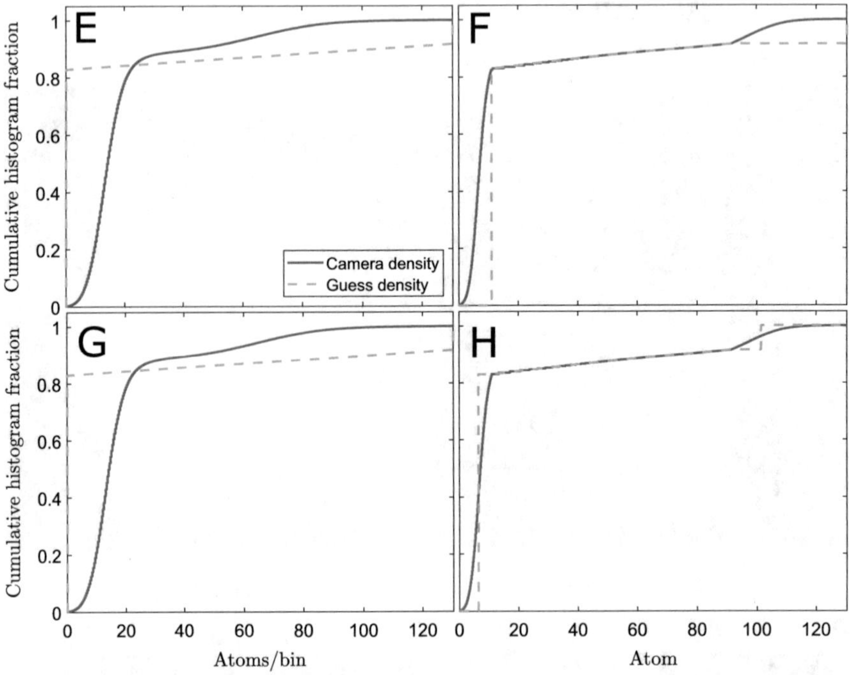

Fig. 4.9 (continued)

When specifying the density target thought to be the best representation of the density image taken by the camera for a given DMD pattern, we usually know approximately what the spatial profile will look like from the Thomas-Fermi predicted density for the projected potential. We however do not usually have as good of an approximation of the amplitude of the signal because it requires knowing the number of atoms for a given shot (fluctuates ~5% shot-to-shot), the intensity profile of the probe beam (drifts over the day), the efficiency of the image collection system, the aberrations in the imaging system, etc. For these reasons, it tends to be impractical to specify the expected density pattern more precisely than simply with the expected relative density profile (Fig. 4.9B). To recognize the images mathematically, we require that the two images be on the same scale so that they can be fairly compared. To achieve this, we first renormalise the range of our guess density to the range of the image we are trying to compare it to and then we apply midway equalization [44, 45] between the two images. This leads to the transformation of Fig. 4.9A, B before equalization into Fig. 4.9C, D after equalization, with Fig. 4.9G showing the cumulative histogram before midway equalization but after data range normalization and Fig. 4.9H showing results after equalization. The equalization is performed using Algorithm 1. Figure 4.9E, F are the same as Fig. 4.9G, H, but using the algorithm presented in [46] to perform the normalization. The main difference between the two algorithms is that ours normalizes to the weighted mean, whereas the scheme used in [46] normalizes to the bin with the closest count value, which is problematic for images with a large number of pixels with exactly the same value. The advantage is apparent when looking at the guess density in Fig. 4.9H which crosses the camera image density at the center of mass of the rises in density, whereas in Fig. 4.9F it can be seen that the overlap of the images can be further improved by simply shifting the position of the initial rise in the cumulative density of the guess density.

4.5.2 Feedback with a Static Potential

Now that we can create a meaningful mapping between the density images on the camera and the DMD pattern uploaded, we need to decide which type of multi-level patterns to create with the DMD. There are two main methods to create multi-level potentials using a direct imaged DMD. The first and simplest is using a static feedback which relies only on PSF of the system being bigger than the spacing of the mirrors imaged on the atom plane to allow for the creation of multiple levels for the potential. The second method is through time averaging, which consists of turning the mirror on/off quickly enough that the atoms only see the average of the time average of the light potential. The advantage of time averaging is that one has access to more grayscale levels that increase with the number of images averaged together. A drawback is that the time dependence of the potential might cause micro-motion in the system, or lead to heating of the atom cloud (see Fig. 4.6E). In this section, we discuss feedback of halftoned static potentials, while in the next one we discuss the use of time averaging. The feedback procedure is shown followed by example feedback potentials and their potential applications.

Before feeding back using the density, the target potential and associated density pattern need to be determined. The conversion from potential to density can be done using a Thomas-Fermi approximation (see Sect. 2.2.4). This density pattern is then used as the target the feedback tries to achieve. Figure 4.10 depicts the main components of a feedback sequence to generate a uniform circular reservoir with uniform flat potential, surrounded by a potential with a linear gradient going from 0 to μ over 360°. The expected density pattern $n_T(i, j)$ for this potential is shown in Fig. 4.10A. Another step to perform before feeding back is to determine the region one wishes to feedback on. In this case, we are only interested in correcting the gradient part of the density target so our feedback region ($\mathscr{R}$) is as shown in Fig. 4.10B. Lastly, the initial DMD pattern needs to be determined. To do so we use a Floyd-Steinberg algorithm [34] modified to account for the incident beam intensity profile, which is to produce Fig. 4.10C. As with any numerical iteration, the closer the initial guess is to the optimal pattern, the faster the optimization converges, the less likely it is to diverge, or converge to a local minimum instead of the global minimum.

To initialise the feedback process, the DMD guess pattern is first used, and the atoms are condensed into the static potential. It is important to condense into the potential instead of ramping on the potential after evaporation since Thomas-Fermi

Data: two discrete images $u_p(i, j)$ of dimensions $(i, j) \in \{1, \dots, N_p\} \times \{1, \dots, M_p\}$, $p \in \{0, 1\}$;
two normalization sets $n_p(l)$ which can be subsets of the images u_p or all of it, $l \in \{1, \dots, L_p\}$;
number of bins over which to equalize N
Result: two midway normalized images $\hat{u}_p(i, j)$
// Normalizing all the data to the wanted range ($\in \{0, \dots, r\}$)
$r = \max(n_1(l)) - \min(n_1(l))$
$dr = r/N$ // Histogram bin size
$u_p(i, j) = (u_p(i, j) - \min(n_p(i, j)))/(\max(n_p(i, j)) - \min(n_p(i, j))) * r$
$n_p(i, j) = (n_p(i, j) - \min(n_p(i, j)))/(\max(n_p(i, j)) - \min(n_p(i, j))) * r$
for $p \in \{0, 1\}$ **do**
  // Computing cumulative histograms H_p of n_p **for** $l \in \{1, \dots, L_p\}$ **do**
    index = $\max(\lceil n_p(l)/dr \rceil, 1)$
    $H_p(\text{index}) = H_p(\text{index}) + 1/L_p$
  end
  for $k \in \{1, \dots, N\}$ **do**
    $H_p(k) = H_p(k) + H_p(k-1)$
  end
end
// Computing the transform function f_p for $u_p(i, j) \to \hat{u}_p(i, j)$
$l = 0$;
for $p \in \{0, 1\}$ **do**
  for $k \in \{1, \dots, N\}$ **do**
    w = 0;
    s = 0;
    while $H_p(k) > H_{\bar{p}}(l)$ **do**
      $l = l + 1$;
      w = w $+\left[H_{\bar{p}}(l) - H_{\bar{p}}(l-1)\right](l - 1/2)dr$
      s = s $+ H_{\bar{p}}(l) - H_{\bar{p}}(l-1)$
    end
    if $s == 0$ **then**
      s = 1;
      w = $(l - 1/2)dr$
    else
    end
    $f_p(k) = 1/2((k - 1/2)dr +$ w/s)
  end
  // Applying the contrast f_p to go from $u_p(i, j) \to \hat{u}_p(i, j)$
  for $(i, j) \in \{1, \dots, N_p\} \times \{1, \dots, M_p\}$ **do**
    $\hat{u}_p(i, j) = f_p(\max(\min(\lceil u_p(l)/dr \rceil, N), 1))$
  end
end

Algorithm 1: Weighted midway equalization

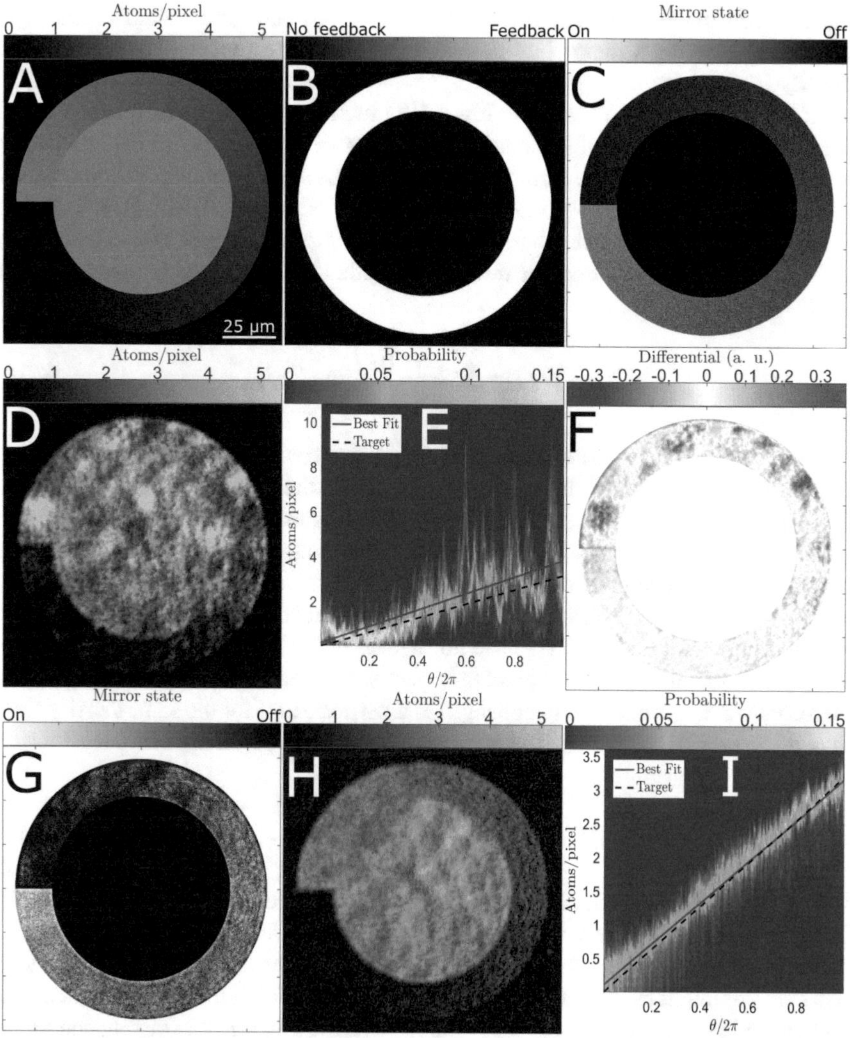

Fig. 4.10 Static DMD feedback example. **A** is the target density. **B** is the specified feedback region which is the only region of the DMD which will be modified during the iterative process. **C** is the guess DMD pattern to approximate the potential leading to **A** generated using the Floyd-Steinberg algorithm [34] modified to account for the incident light beam profile. **D** is the average of three density images taken using the same static DMD pattern **C** it is the starting point of the feedback. **E** is the angular decomposition of **D** in the feedback region (**B**), in the solid red line shows the best fit to the data and the dashed black line is the angular decomposition of the target (**A**). **F** shows the error map which is the difference between **B** and **A** scaled by the local light intensity. **G** shows the feedback DMD pattern after 10 iterations, with **H** being the density profile resulting from averaging three runs with **G** as the DMD pattern. **I** is the same angular decomposition as **E** but for **H** instead of **D**

(assumed during feedback) requires that the atoms be in the groundstate which might not be the case for a ramped on potential. To lower the shot-to-shot noise on the density images, we usually average 2–5 runs of the same pattern together giving a density image $n_M(i, j)$, akin to Fig. 4.10D which was transformed to the DMD frame using the inverse of the image recognition determined coordinate transform between the DMD and the camera. Figure 4.10E shows the angular decomposition of the density image Fig. 4.10D (histogram) and target Fig. 4.10A (black dashed line) in the specified feedback region Fig. 4.10B after the target and density have been normalized to one another through normalization of the mean counts on the feedback region

$$\hat{n}_T(i, j) = \frac{\sum_{\mathscr{R}} n_M(i, j)}{\sum_{\mathscr{R}} n_T(i, j)} n_T(i, j). \tag{4.10}$$

From the normalized density and target density, an error map can be calculated which reflects the local error on the image by subtracting the target from the density and dividing the result by the highest density. The highest density can be thought of as the location where the overall potential is minimum and normalises the density range for a given chemical potential. The result is divided by the local incident light potential $V(i, j)$ (see Sect. 4.6), in units of the chemical potential at the given location assuming the DMD is fully 'on'. We normalize by the local intensity as changes in the number of mirrors have a greater impact where the incident light is high. One therefore wants to make smaller changes in those locations for the same density error than in a lower light regions. The error map is therefore given by

$$E(i, j) = \alpha\mu(n_M(i, j) - \hat{n}_T(i, j))/(\max(n_M(i, j)) - \min(n_M(i, j)))/V(i, j) \tag{4.11}$$

where $V(i, j)$ is the potential depth associated with illumination intensity $I(i, j)$. The parameter α is a numerical parameter specifying the step size and is used to control how quickly the algorithm tries to converge. If it is too large then oscillations will occur or might even lead to divergence of the result. If it is too small, the feedback will take a long time to converge. To address this, the step size can be reduced as the feedback converges, but here we simply keep it constant.

Figure 4.10F shows the error function for Fig. 4.10A, D using $\alpha = 1$. The DMD pattern is then fed back using the error map to determine the probability of switching the mirrors. This is implemented by generating random numbers between 0 and 1, on the same grid as the error map. If the random number is lower than the local error, the mirror is switched. Before switching the state of the mirror, we check that switch the mirror lowers the error by checking the state against the sign. The reason for this check is to avoid turning off (on) mirrors to correct for a density that is too high (low). If the mirror-test returns a positive for these two conditions, the mirror is flipped. As the algorithm converges the amplitude of the error signal decreases, effectively taking smaller and smaller steps until it converges to the final solution. Figure 4.10G shows the optimized DMD pattern after 10 iterations, with Fig. 4.10H showing the resulting density pattern, and Fig. 4.10I showing the angular decomposition of density in the feedback region of Fig. 4.10B. There is a significant improvement over the initial guess which gave Fig. 4.10D, E. The root means square variation between the measured density and the best fit linear gradient density improves from 52% to 18% after the feed-forward is performed. The code is not deterministic in its optimization but this is actually an advantage as it allows the exploration of more of the parameter space. Also, if in doubt that it has hit a local minimum one can re-run the optimization a second time, although we have not encountered any scenario where the program would not converge.

4.5.3 Example of a Fed-Back Static Potential

Using this feedback method, one can generate the patterns shown in Fig. 4.11. The first pattern (Fig. 4.11A) consists of a uniform density ring surrounded by a density gradient created by a linear potential gradient around the ring. This fed-back gradient is used to phase imprint [Eq. (2.65)] persistent currents on the rings as proposed in [47]; see Sect. 4.5.4 for a detailed description. A hard walled density box is also generated (Fig. 4.11B) which consists of a surrounding box of high-density fluid with a centre shell of low-density fluid. This geometry can be used as a vortex trap, as there is an energy barrier for vortices moving from a low density to a high density medium, since this requires moving more mass at the same velocity in the high density region (see Eq. (2.47)). However, it was found that the feedback effects the dynamics of the vortices and determining the mechanism is the subject of ongoing study. A likely explanation is that there are features on the order of the healing length which cannot be imaged by the camera, but that appear during feedback and effect the vortex dynamics.

We also create a simple density step consisting of one region of low density connected to a region of high density Fig. 4.11C. This will be used in tandem with deterministic vortex creation [48], which is easily realizable with the dynamic property of the DMD (see Fig. 4.12), to study vortex dipole optics, a subject that has been proposed and numerically studied by the group of Dr Ashton Bradley at the University of Otago [49]. An illustration of the basic principle of vortex optics is shown in Fig. 4.13. Figure 4.11D demonstrates a feedback image of Bose and Einstein in a BEC for illustrative purposes.

4.5.4 Persistent Current Imprinting

Current imprinting is done using a linear light intensity gradient along with a barrier. The first step is to create the linear gradient, and this is done using feedback, with the results shown in Fig. 4.11A. A circle BEC of uniform density atoms is then created using a hard wall potential with radius r_1. Next, a barrier is ramped on at r_0 over

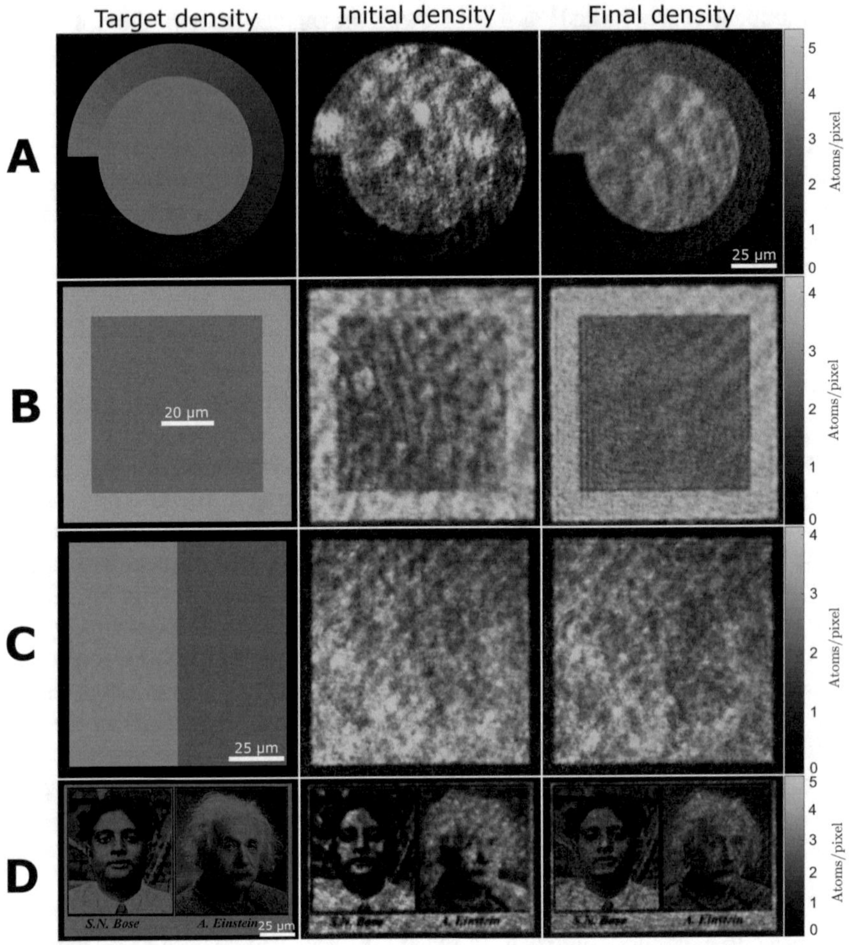

Fig. 4.11 Examples of halftoned feedback images showing the target density, the initial density pattern, and density pattern after 10 iterations of the feedback algorithm. Density images are made up of 3 camera images of the density averaged together. **A** shows the creation of a linear gradient, **B** of a density box, **C** of a reservoir with a density step, and **D** of a grayscale picture of Bose and Einstein

100 ms, while also ramping on a barrier, in the outer ring, at ϕ_0 as shown in Fig. 4.14A. The atoms are not condensed in the ring and barrier geometry because it leads to a non-uniform density due to the difference in the local density of the thermal cloud overlapping the outer ring and centre circle before condensation.

To start the imprinting process, the linear intensity gradient is turned on and the barrier's angular position is accelerated at the same rate and the same direction as the atoms are accelerated by the gradient. The angular acceleration is given by $a_0 = V_0/(m(2\pi\bar{r})^2)$, with m being the ^{87}Rb atomic mass, $\bar{r} = (r_0 + r_1)/2$ and μ the

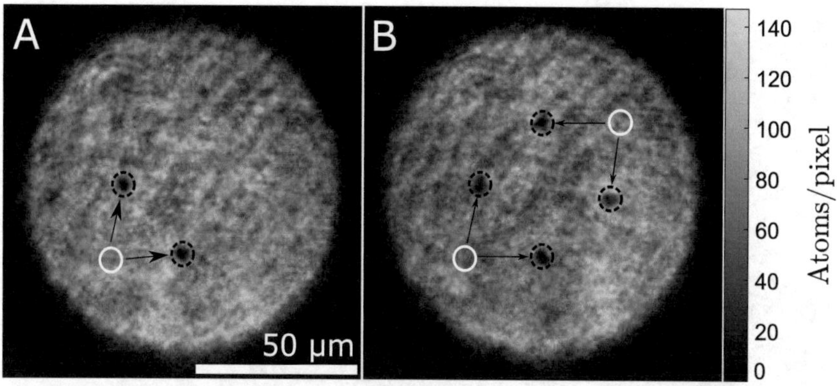

Fig. 4.12 Using a DMD to create vortex dipoles with the *chopsticks* method demonstrated in Ref. [48]. **A** shows the creation of one vortex dipole pair and **B** the creation of two vortex dipole pairs. The white circle shows the initial location of the pinning beam which splits into two travelling pining beams that travel in a straight line along the arrows, towards the locations circled in black. In both cases, the chopsticks beams are travelling at 60° from one another, for 25 μm, at a velocity of 230 μm/s (∼0.2c, where c is the speed of sound)

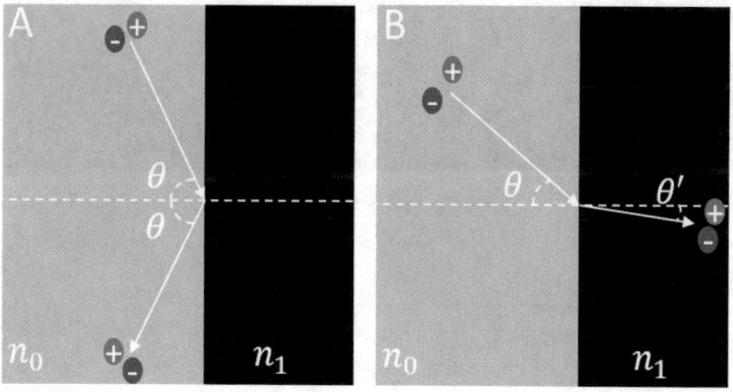

Fig. 4.13 Vortex dipole optics uses vortex dipoles instead of photons to interact with an interface. The interface is characterized by a change in the density, instead of in the refractive index. Illustrated are a reflection (**A**) and a transmission (**B**) of a vortex dipole pair at the interface which obeys a similar law to Snell's law in the case of the transmission. To conserve the energy with $n_1 > n_0$ in **B** the vortex dipole separation distance changes after travelling through the interface. The analogues between optics and vortex dipoles go further than Snell's law type relationship. It is possible for the interface to trap the dipole pair. The pair can penetrate the high refractive index during reflection, similar to the evanescent waves in optics

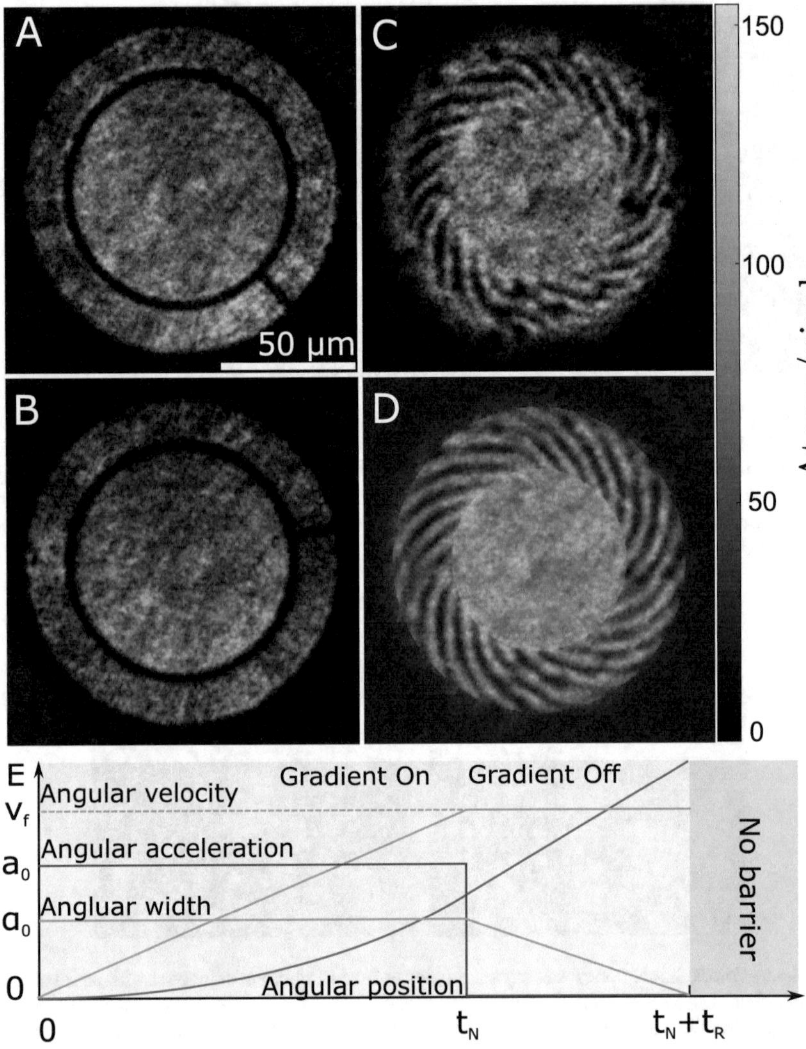

Fig. 4.14 Current imprinting using light gradient. **A** shows the phase imprint phase with the linear gradient on (not visible) and the barrier accelerated at the same rate as the atoms. **B** is the barrier ramp down in which the linear gradient is turned off and the barrier is rotated at a constant velocity while being ramped down. **C** TOF release of the cloud 100 ms after the barrier ramps down. **D** is a fit to **C** to determine the number of imprinted quanta of circulation using $n(r, \phi) = A\Theta(r_0 - r) + B\,[1 + \sin(N\phi + Cr + \phi_0)]\,\Theta(r - r_0)\Theta(r_1 - r)$ where $\Theta(x)$ is the Heaviside function with resulting N = 22. **E** presents the angular width, position, velocity, and acceleration of the barrier during the sequence. The barrier initially starts stationary at position 0 and is accelerated with a_0 with a constant width while the linear gradient is on (orange box). Once the gradient is turned off (purple box), the barrier is rotated at a constant velocity and the width the barrier is linearly ramped down until the barrier is gone

height of the gradient. For this case $V_0 = \mu$, where μ is the chemical potential of Fig. 4.11A after the feedback. This is obtained by using Eqs. (2.65) and (2.30) with the linear gradient potential $V(\phi) = V_0(\phi - \phi_0)/2\pi$. The linear gradient is held on and the barrier is accelerated until the desired quanta of circulation, N, is imprinted. This takes a time $t_N = 2N\pi\hbar/\mu$ which can be found by solving

$$\theta(\phi + \pi, t) - \theta(\phi - \pi, t) + 2N\pi = \theta(\phi + \pi, t + t_N) - \theta(\phi - \pi, t + t_N), \tag{4.12}$$

where θ is the phase of the wavefunction given by Eq. (2.65). The linear gradient is then turned off and the barrier is rotated at constant angular velocity $v_f = a_\phi t_N$ over 90° shown in Fig. 4.11B. During this rotation, the angular width of the barrier is ramped down from α_0 to 0. The time of this step is $t_R = \pi/2a_\phi t_N$. Once the barrier is gone, a N quanta circulation should have been established. The BEC is then allowed to evolve freely for a period of 100 ms. The cloud is then released for a short time-of-flight (15 ms) during which the outer ring interferes with the stationary inner condensate. The number of azimuthal interference fringes corresponds to the imprinted circulation. The interference pattern is shown in Fig. 4.11C, where we also notice vortices in the outer ring. These are speculated to come from non-uniformity in the radial profile of the linear intensity gradient, causing nucleation of vortices.

The interference pattern can be approximated by

$$n(r, \phi) = A\Theta(r_0 - r) + B\left[1 + \sin(N\phi + Cr + \phi_0)\right]\Theta(r - r_0)\Theta(r_1 - r) \tag{4.13}$$

where $\Theta(x)$ is the Heaviside function and A, B, N, C, and ϕ_0 are free parameters. In this case, one finds that the imprinted current has a circulation of 22 quanta by counting the azimuthal fringes as done in Fig. 4.11D. The angular position, velocity, acceleration and width of the barrier in the outer ring as a function of time are shown in Fig. 4.11E. Note that the only reason the barrier is required is to break the phase matching condition of the superfluid. Without the barrier, a linear gradient would not lead to a circulation in the ring, as the phase will smoothly reconnect.

Another method for generating currents is to perform a similar procedure but without the linear gradient. In this case, there is no direct imprinting of the phase except at the barrier wall where the imprinted phase gradient is steep. Because it takes time for the system to react to the imprinted phase near the barrier (the method can also be thought of as the barrier pushing the fluid) the angular acceleration of the barrier must be slower than in the imprinted case, $a_0 \ll V_0/(m(2\pi\bar{r})^2)$, which makes the ramping on of the current much longer than in the case of imprinting with a gradient in the same system for the same circulation. Barrier acceleration that is too fast results in extraneous excitations. The results are shown in Fig. 4.15. The barrier ramp slowly accelerates the barrier until it is rotating at the angular velocity that matches a N circulation vortex for a given ring radius, $v_F = N\hbar/(2\pi\bar{r})^2 m$. Subsequently, the procedure develops similarly, with a ramp down of the barrier followed by an equilibration time, and finally a TOF step to observe the result of

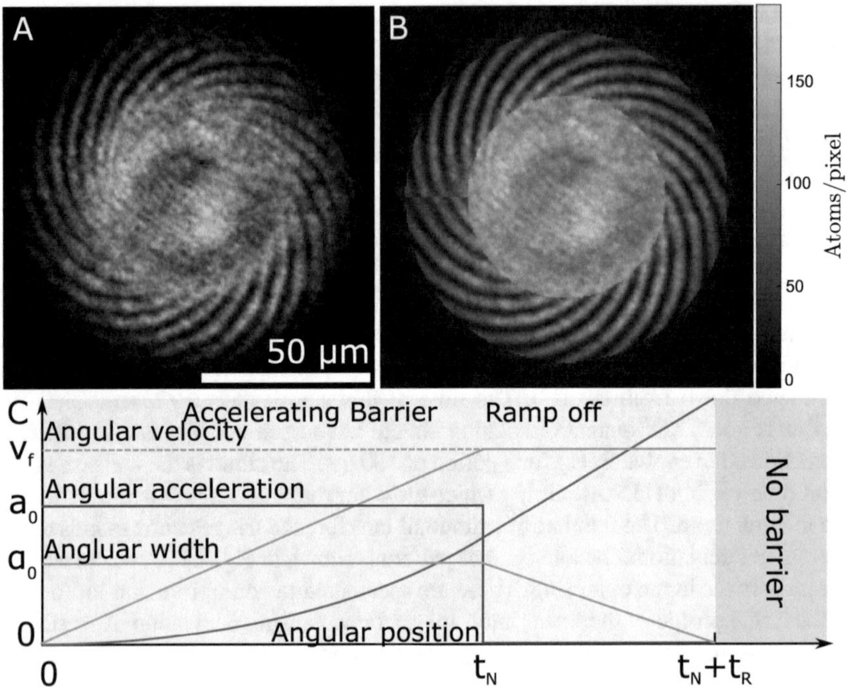

Fig. 4.15 Perpetual current generation using stirring. **A** shows the interference between the outer ring and the inner ring after a current was generated in the outer ring using a barrier. **B** shows the overlap between **A** and a fit to determine the number of imprinted quanta of circulation. The functional form used is $n(r, \phi) = A\Theta(r_0 - r) + B\,[1 + \sin(N\phi + Cr + \phi_0)]\,\Theta(r - r_0)\Theta(r_1 - r)$ where $\Theta(x)$ is the Heaviside function with resulting N = 22

the imprinting. The position and ramp down sequence of the barrier are shown in Fig. 4.15C, with Fig. 4.15A showing the resulting interference pattern and Fig. 4.15B showing a best fit with a circulation of 22. The data is quite clean and there are no extra excitations when compared with Fig. 4.14C. However, the gradient imprint method can in principle generate the same current faster without causing excitations. The imprint time used here to imprint 22 quanta of circulation in the linear gradient case was 44 ms whereas 600 ms was used for stirring with the barrier.

4.5.5 Time-Averaged Feedback

Time-averaged potentials are more complicated than their static counterparts due to their dynamics, but come with the advantage of allowing finer control of the potential. This results from decreased spacing between the grayscale levels by a factor of N_{avg}, where N_{avg} is the number of DMD images averaged together. Here, our preliminary method for feeding back a time-averaged potential is introduced. The method is closely related to the static pattern feedback presented in the previous section.

To keep track of the DMD patterns to be time averaged, an approach would be to create a 3D matrix, with 2-dimensions representing the spatial dimension of the DMD, and the third representing time-axis with length N_{avg}. From there, it is possible to follow the steps used for the static feedback by applying the error map to each time layer. This method indeed improves the density pattern with each iteration and converges towards the target, the drawback is that the temporal spacing between the mirrors is not optimized to minimize the variance between the potentials. In other words, each of the potentials that makes up the time-averaged potential should be as similar as possible to minimize heating and micro-motion from the varying potential. Instead, in order to minimize the variance and simplify the feedback, is to collapse the 3D matrix DMD state into a 2D matrix $P(x, y)$ where each entry determines the expected duty cycle of the 'off' mirror during a time average cycle. One can feedback on this map by increasing or decreasing the local duty cycle based on the error in the local measured error in the density. The DMD patterns for the different time steps are then generated while performing the following minimization

$$\begin{aligned} \underset{E}{\text{minimize}} \quad & E = \frac{1}{2}\sum_{i=1}^{N_{\text{avg}}}\sum_{j=1}^{N_{\text{avg}}}\iint dxdy[(DMD(x,y,t_i) - DMD(x,y,t_j))I(x,y) \circledast PSF(x,y)]^2, \\ \text{subject to} \quad & \sum_{i=1}^{N_{\text{avg}}} DMD(x,y,t_i) = M(x,y). \end{aligned} \tag{4.14}$$

$DMD(x, y, t) \in \{0, 1\}$ is the state of the DMD mirrors at position x, y and time t. $I(x, y)$ is the local illumination intensity and $PSF(x, y)$ is the point spread function of the DMD imaging optics. $M(x, y) \in \{1, \ldots, N_{\text{avg}}\}$ is a matrix storing the number of mirrors that need to be on for each pixel of the DMD, over the time averaging cycle, and is calculated using

$$M(x, y) = \lfloor P(x, y)N_{\text{avg}} \rfloor + [(P(x, y)N_{\text{avg}} - \lfloor P(x, y)N_{\text{avg}} \rfloor) > rand], \tag{4.15}$$

where $rand$ is a generated random number between 0 and 1 with uniform distribution. Since not all duty cycles will lead to an integer number of mirrors on for each spatial location, the residual is used, as the probability of increasing the total number of mirrors off by one at each location.

Algorithm 2 creates DMD patterns for time averaging that minimize the variance between the frames. The presented algorithm is a minimum working example; it can be greatly sped up by keeping track of the convolution as mirrors are added instead of performing a convolution each time.

Figure 4.16A compares Algorithm 2 to a random method for a random duty cycle map. The algorithm performs better than the random assignment method by about 2.5

order of magnitude, see Fig. 4.16B. Note that a solution to Eq. (4.14) is not unique, and a further constraint that could be added is the maximization of the temporal displacement between the mirrors at a given physical location. The minimum number of solutions would be N_{avg} solutions which are simply a temporal displacement of the same solution, which is equivalent due to the cyclical nature of time averaging.

Data: discrete duty cycle map $P(i, j) \in [0, 1]$ of dimensions
$(i, j) \in \{1, \ldots, N_p\} \times \{1, \ldots, M_p\}$,
the number of time averaged levels N_{avg},
local DMD illumination intensity $I(i, j)$,
and point spread function of the optical system $PSF(i, j)$
Result: DMD patterns to be used for the time averaging $DMD(i, j, k)$, $k \in \{1, \ldots, N_{avg}\}$

```
// Creating local number of 'off' mirror in range {1, ..., N_avg}
M(x, y) = ⌊P(x, y)N_avg⌋ + [(P(x, y)N_avg − ⌊P(x, y)N_avg⌋) > rand]
// Initializing the images
DMD(i, j, k) = false([dmdSize(2),dmdSize(1),numLevels]);
for i ∈ {1, ..., N_p} do
    for j ∈ {1, ..., M_p} do
        // Calculate the light level at location i,j based on the currently activate mirrors for
        k ∈ {1, ..., M_p} do
            L(k,:,:) = ifft2(fft2(DMD(:,:,k)*I(:,:))*fft2(PSF(:,:)));
        end
        // Randomize the vector index then sort in ascending order and flip mirrors where the
        light level is smallest
        indexRand = randperm(length(L(:,i,j))) [ ,index] = sort(L(indexRand,i,j),'ascend')
        DMD(indexRand(index(1:N_avg)),i,j) = 1;
    end
end
```

Algorithm 2: Variance minimization for time averaged DMD pattern generation

Before feeding back on the duty cycle map using the density, the desired potential and associated density pattern need to be determined. The conversion from potential to density can be done using a Thomas-Fermi approximation (see Sect. 2.2.4). This density pattern is then used as the target for the feedback. Figure 4.17 depicts the main components of a feedback sequence to generate a two-levels density box (Note the steps are similar to the static feedback case). The resulting density target is depicted in Fig. 4.17A. Figure 4.17B specifies the region that needs to be fed back and Fig. 4.17C shows the initial guess for the duty cycle map, which is simply a linearly scaled version of Fig. 4.17A.

During the feedback process, the guess duty cycle is used to create the DMD patterns and uploaded to the DMD. The atoms are then condensed before the time-averaged pattern is turned on, and left to evolve for one second so that they redistribute to their ground state. Note that condensing in the time-averaged potential would be better, but it seems very detrimental to the condensation process which is currently not understood, since no heating should be occurring as per Fig. 4.6E for averaging above >3 kHz. To lower the shot to shot noise on the density images, we usually

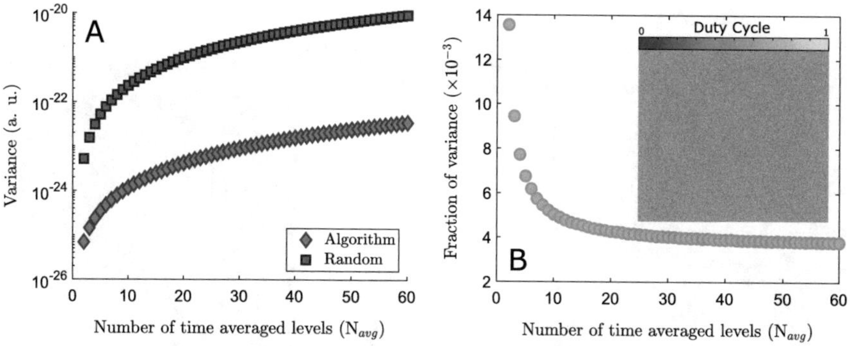

Fig. 4.16 Numerical comparison of Algorithm 2 with simple random generation for the generation of the time-averaged DMD potential. **A** shows the variance of the potentials calculated using Eq. (4.14a) versus the number of time-averaged levels for a random duty cycle map, $P(x, y)$, shown in the inset of **B**. **B** shows the ratio of the variance of the algorithm pattern to that of the random generated pattern. It can be seen that the variance of random temporal assignment of mirrors is about 2.5 orders of magnitude worse

average 2–5 runs of the same pattern together giving a density image $n_M(i, j)$ akin to Fig. 4.17D, which was transformed to the DMD frame using the inverse of the image recognition determined coordinate transform between the DMD and the camera. Figure 4.17E shows the projection onto the y-axis of Fig. 4.17D as a 2D histogram, the projected target Fig. 4.17A (black dashed line) mean normalize using Eq. (4.10) and the best fit to the data (red solid line) are also shown. Equation (4.11) is used to calculate the error map between Fig. 4.17A, D which is shown in Fig. 4.17F for $\alpha = 1$. The DMD pattern is then fed back using the error map as a probability of switching the mirrors. We correct the duty cycle map using

$$P'_{n+1}(x, y) = \max(\min(P_n(x, y) - E(x, y), 1), 0), \tag{4.16a}$$

$$P_{n+1}(x, y) = P'_{n+1}(x, y)/\max(P'_{n+1}(x, y) - 1). \tag{4.16b}$$

The first step does the correction and guarantees that the range is between 0 and 1. The second step makes sure that at least one of the mirrors is always off. Figure 4.10G shows the optimized duty cycle map after 20 iterations. The resulting density pattern is shown in Fig. 4.10H, I shows the same projection as Fig. 4.17E, but for projection of Fig. 4.10H onto the y-axis. As expected of the feedback, there is a significant improvement over the initial guess which gave Fig. 4.10D, E.

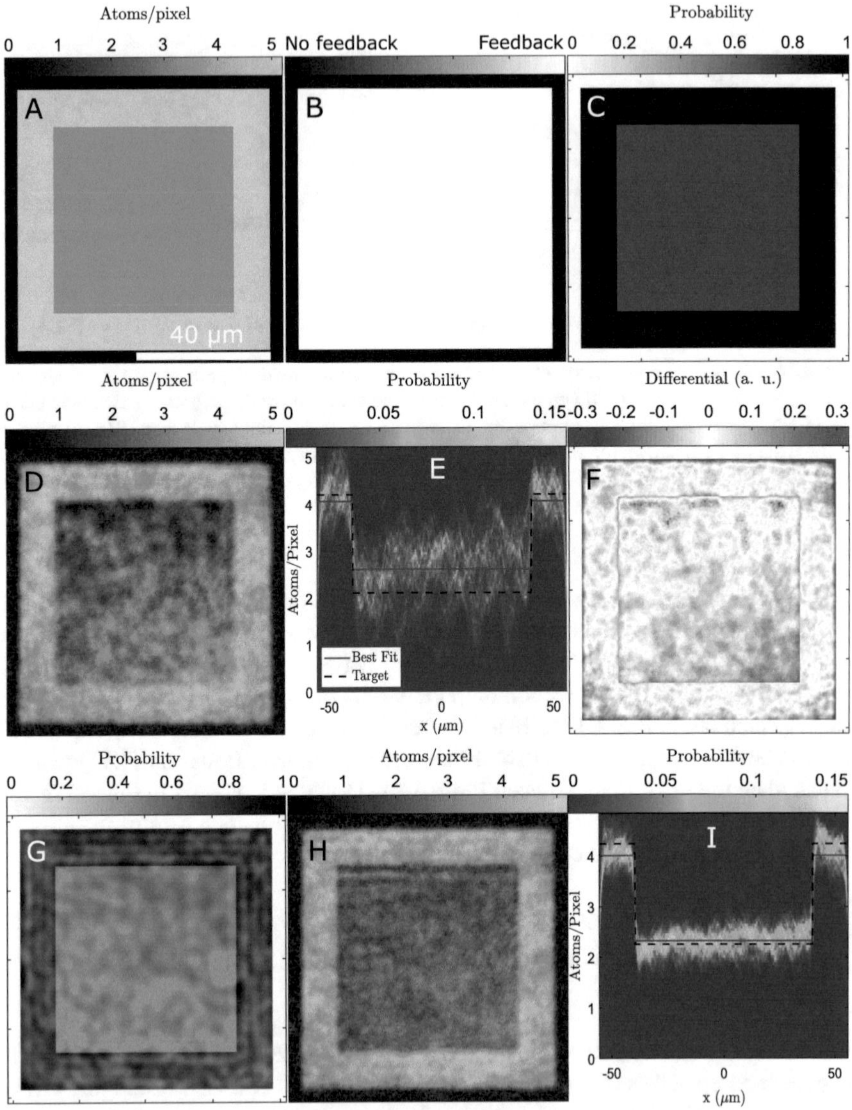

Fig. 4.17 Time-averaged DMD feedback example using $N_{avg} = 3$ levels. **A** shows the target density of the feedback. **B** is the specified feedback region which is the only region of the DMD which will be modified during our iterative process. **C** is the guess duty cycle map to approximate the potential leading to **A** atomic distribution. **D** is the average of three density images taken using the same time averaged DMD pattern **C** at different hold times $\{250, 750, 1250\}$ ms. **E** is the histogram projection of **D** onto the y-axis of the central 400 pixels, the solid red line shows the best fit to the data and the dashed black line is the same projection for the target **A**. **F** shows the error map which is the difference between B and A scaled by the local light intensity using Eq. (4.11). **G** shows the feedback DMD pattern after 20 iterations, with **H** being the same as **D** but for the duty cycle map (**G**). **I** is the same angular decomposition as **E** but for **H** instead of **D**

4.6 Modelling the Potential

In most of the experiments, the atoms are trapped in an optical dipole sheet in the vertical direction (V_{sheet}), a DMD projected potential provides the trapping in the horizontal direction (V_{DMD}) and a levitation field (V_{mag}) is used to cancel gravity (V_{gravity}). The total potential seen by the atoms is the sum of all the constituent potentials

$$V(\mathbf{r}, t) = V_{\text{sheet}}(\mathbf{r}) + V_{\text{DMD}}(\mathbf{r}, t) + V_{\text{mag}}(\mathbf{r}) + V_{\text{gravity}}(\mathbf{r}). \tag{4.17}$$

The vertical sheet confinement comes from an elliptical Gaussian beam with a vertically focussed waist of ($w_0^z \sim 8.6$ µm) and a collimated horizontal waist of ($w_0^x \sim 500$ µm). The sheet trapping potential is thus given by

$$V_{\text{sheet}}(\mathbf{r}) = \frac{2\gamma_{1064} P_{\text{sheet}}}{\pi w^x(y) w^z(y)} \exp\left[-2\left(\frac{z}{w^z(y)}\right)^2 + \left(\frac{x}{w^x(y)}\right)^2\right], \tag{4.18}$$

where P_{sheet} is the beam power in the atom plane, γ_{1064} is the conversion factor for going from laser light intensity to trapping potential, Eq. (2.63), and

$$w^i(y) = w_0^i \sqrt{1 + \left[\frac{\lambda y}{\pi (w_0^i)^2}\right]^2}, i \in \{x, z\}, \tag{4.19}$$

is the waist, dependent on the displacement from the focal point, for a beam travelling in the y-direction.

To calculate the DMD potential, we start from the DMD illumination intensity pattern, which has waist $w^y = w^x = 9.81$ mm at the DMD plane, and a 98.1 µm waist when projected onto the atom plane,

$$I_{\text{DMD}}(x, y) = \frac{2 P_{\text{DMD}}}{\pi w^x w^y} \exp\left[-2\left(\frac{y}{w^y}\right)^2 + \left(\frac{x}{w^x}\right)^2\right], \tag{4.20}$$

where P_{DMD} is the beam power that would make it to the atom plane if the DMD was not masking the edges of the beam. To get the potential at the atoms Eq. (4.20) is first multiplied by the intensity by the DMD pattern $DMD(x, y)$, which is a binary mask of either 0 and 1, before convolving the resulting pattern with the PSF of the optical system and converting to energy from intensity using Eq. (2.63). The final potential is thus given by

$$V_{\text{DMD}}(\mathbf{r}, t) = \gamma_{532} DMD\left(\frac{x+y}{\sqrt{2}}, \frac{y-x}{\sqrt{2}}, t\right) I_{\text{DMD}}\left(\frac{x+y}{\sqrt{2}}, \frac{y-x}{\sqrt{2}}\right) \circledast PSF(x, y) \tag{4.21}$$

where the rotation of the DMD when projected onto the atom plane is accounted for, but the z-dependence of the illumination light and of the PSF is not considered.

The magnetic field potential is modelled using Eqs. (3.3) and (3.1) as is done in Sect. 3.3.1. The gravitational potential is given by

$$V_{\text{gravity}}(\mathbf{r}) = m_{\text{Rb}}gz. \tag{4.22}$$

Unless stated otherwise, this model is used for the purpose of modelling the potential throughout the numerical simulations described in this thesis.

4.7 Conclusion

DMD devices, combined with commercially-available glass-corrected objectives, present a powerful technique for both the microscopic patterning of quantum gases. By utilising a commercial fluorescence cell with relatively thin 1.25 mm walls, expensive custom-built objectives can be avoided while still achieving high resolution. We are able to estimate a 630(10) nm FWHM PSF for our DMD projection system at 532 nm illumination, in agreement with a single-mirror analysis. A similar estimate of the PSF at 780 nm using a single mirror gives 960(80) nm FWHM, while MTF analysis of an atomic resolution target gives 1250(20) nm FWHM. Some of the broadening is due to diffusion and using Faraday imaging gives a PSF FWHM of 1080(15) nm.

Commercial DMD devices are now well-suited for implementation in both quantum gas experiments and other optical trapping applications. The painted-potentials technique, in combination with error diffusion methods, can generate grayscale potentials with more than one hundred levels. The ability to easily store and trigger 1,700 frames at high speed shows great promise for the dynamic control of these microscopic potentials. Furthermore, the ability to nearly instantaneously quench the potential geometry has many potential applications, such as investigations of superfluid transport between reservoirs [50, 51], and a demonstration of the superfluid fountain effect in BECs [52]. These techniques have proven usefull for the creation of deterministic superfluid turbulence in BECs in the production and control of vortex pairs using moving potential barriers similar to what was done in Refs. [48, 53]. Configurable trap geometries are also usefull for designing atomtronic circuits [54] such as dumbbell as is done in Chap. 5. As the DMD is largely wavelength insensitive (except in terms of diffraction efficiency), the possibilities for use at multiple wavelengths are intriguing. This capability could be utilised for the production of species dependent potentials [55].

While direct imaging is powerful and very versatile, directly using it to create halftoned images leads to any imperfection in the imaging system and illumination light being imprinted onto the projected pattern. We presented a direct density feedback method that can minimize the imperfection in the projected pattern. This technique is useful for any application that requires a non-uniform potential, such as phase imprinting, vortex dipole optics and traps, and artistic shape imaging.

More generally, our results demonstrate the great utility of direct imaging of an DMD. With a well-corrected optical system, performance close to the diffraction limit can still be achieved at a non-trivial 0.45 NA. These results are also applicable to optical trapping beyond quantum gases, and in particular may be advantageous for the production of numerous traps for confining arrays of particles [4, 13, 14]. The flexibility of the DMD is used to study superfluid transport between reservoirs in Chap. 5 and the full dynamics capability of the DMD is used in the generation of negative-temperature Onsager vortex clusters in Chap. 6.

References

1. Grier DG (2003) A revolution in optical manipulation. Nature 424:810
2. Woerdemann M, Alpmann C, Esseling M, Denz C (2013) Advanced optical trapping by complex beam shaping. Laser Photonics Rev 7:839–854
3. Neuman KC, Block SM (2004) Optical trapping. Rev Sci Instrum 75:2787–2809
4. Chowdhury S et al (2014) Automated manipulation of biological cells using gripper formations controlled by optical tweezers. IEEE Trans Autom Sci Eng 11:338–347
5. Pasienski M, DeMarco B (2008) A high-accuracy algorithm for designing arbitrary holographic atom traps. Opt Express 16:2176–2190
6. Gaunt AL, Hadzibabic Z (2012) Robust digital holography for ultracold atom trapping. Sci Rep 2:721
7. Harte T, Bruce GD, Keeling J, Cassettari D (2014) Conjugate gradient minimisation approach to generating holographic traps for ultracold atoms. Opt Express 22:26548–26558
8. Nogrette F et al (2014) Single-atom trapping in holographic 2D arrays of microtraps with arbitrary geometries. Phys Rev X 4:021034
9. Preiss PM et al (2015) Strongly correlated quantum walks in optical lattices. Science 347:1229–1233
10. Fukuhara T et al (2013) Quantum dynamics of a mobile spin impurity. Nat Phys 9:235–241
11. Zupancic P et al (2016) Ultra-precise holographic beam shaping for microscopic quantum control. Opt Express 24:13881–13893
12. Mogensen PC, Glüuckstad J (2000) Dynamic array generation and pattern formation for optical tweezers. Opt Commun 175:75–81
13. Eriksen RL, Mogensen PC, Glüuckstad J (2002) Multiple-beam optical tweezers generated by the generalized phase-contrast method. Opt Lett 27:267–269
14. Curtis JE, Koss BA, Grier DG (2002) Dynamic holographic optical tweezers. Opt Commun 207:169–175
15. Dudley D, Duncan WM, Slaughter J (2003) Emerging digital micromirror device (DMD) applications. SPIE Proc 4985
16. Liang J, Kohn Jr RN, Becker MF, Heinzen DJ (2009) 1.5% root-mean-square flat-intensity laser beam formed using a binary-amplitude spatial light modulator. Appl Opt 48:955–1962
17. Ha L-C, Clark LW, Parker CV, Anderson BM, Chin C (2015) Roton-maxon excitation spectrum of Bose condensates in a shaken optical lattice. Phys Rev Lett 114:055301
18. Kumar A et al (2016) Minimally destructive, Doppler measurement of a quantized flow in a ringshaped Bose-Einstein condensate. New J Phys 18:025001
19. Stilgoe AB, Kashchuk AV, Preece D, Rubinsztein-Dunlop H (2016) An interpretation and guide to single-pass beam shaping methods using SLMs and DMDs. J Opt 18:065609

20. Mirhosseini M et al (2013) Rapid generation of light beams carrying orbital angular momentum. Opt Express 21:30196–30203
21. Liu R, Li F, Padgett M, Phillips D (2015) Generalized photon sieves: fine control of complex fields with simple pinhole arrays. Optica 2:1028–1036
22. Choi J-Y et al (2016) Exploring the many-body localization transition in two dimensions. Science 352:1547–1552
23. Zimmermann B, Mueller T, Meineke J, Esslinger T, Moritz H (2011) High-resolution imaging of ultracold fermions in microscopically tailored optical potentials. New J Phys 13:043007
24. Schnelle S, Van Ooijen E, Davis M, Heckenberg N, Rubinsztein-Dunlop H (2008) Versatile two-dimensional potentials for ultra-cold atoms. Opt Express 16:1405–1412
25. Henderson K, Ryu C, MacCormick C, Boshier M (2009) Experimental demonstration of painting arbitrary and dynamic potentials for Bose-Einstein condensates. New J Phys 11:043030
26. Bell TA et al (2016) Bose-Einstein condensation in large time-averaged optical ring potentials. New J Phys 18:035003
27. Benjamin Lee (2013) Texas instruments application report
28. Texas Instruments (2012) DLPS025E
29. Texas Instruments (2008) TI DN 2509927
30. Pedrotti FJ, Pedrotti LS (1992) Introduction to optics 2nd. Prentice Hall, United-States
31. Born M, Wolf E (1999) Principles of optics: electromagnetic theory of propagation, interference and diffraction of light, 7th edn. Cambridge University Press, Cambridge
32. Liang J (2012) High-precision laser beam shaping and image projection. Doctor of Philosophy. The University of Texas at Austin, Austin
33. Perego E (2015) Generation of arbitrary optical potentials for atomic physics experiments using a digital micromirror device. Master Degree in Physics. University of Florence, Florence
34. Floyd RW, Steinberg L (1976) An adaptive algorithm for spatial grayscale. Proc Soc Inf Disp 17:75–77
35. Hueck K, Mazurenko A, Luick N, Lompe T, Moritz H (2017) Note: suppression of kHz-frequency switching noise in digital micro-mirror devices. Rev Sci Instrum 88:016103
36. Gupta S, Murch K, Moore K, Purdy T, Stamper-Kurn D (2005) Bose-Einstein condensation in a circular waveguide. Phys Rev Lett 95:143201
37. Eckel S, Jendrzejewski F, Kumar A, Lobb CJ, Campbell GK (2014) Interferometric measurement of the current-phase relationship of a superfluid weak link. Phys Rev X 4:031052
38. Corman L et al (2014) Quench-induced supercurrents in an annular Bose gas. Phys Rev Lett 113:135302
39. Amico L, Osterloh A, Cataliotti F (2005) Quantum many particle systems in ring-shaped optical lattices. Phys Rev Lett 95:063201
40. Mathey L et al (2010) Phase fluctuations in anisotropic Bose-Einstein condensates: from cigars to rings. Phys Rev A 82:033607
41. Horstmeyer R, Heintzmann R, Popescu G, Waller L, Yang C (2016) Standardizing the resolution claims for coherent microscopy. Nat Photonics 10:68–71
42. Muessel W et al (2013) Optimized absorption imaging of mesoscopic atomic clouds. Appl Phys B 113:69–73
43. Mack CA (2006) Field guide to optical lithography. SPIE Press Book, United-States
44. Cox IJ, Roy S, Hingorani SL (1995) Dynamic histogram warping of image pairs for constant image brightness. Proc Int Conf Image Process 2:366–369
45. Delon J (2004) Midway image equalization. J Math Imaging Vis 21:119–134
46. Guillemot T, Delon J (2016) Implementation of the midway image equalization. Image Process Line 6:114–129
47. Kumar A et al (2018) Producing superfluid circulation states using phase imprinting. Phys Rev A 97:043615
48. Samson E, Wilson K, Newman Z, Anderson B (2016) Deterministic creation, pinning, and manipulation of quantized vortices in a Bose-Einstein condensate. Phys Rev A 93:023603
49. Cawte MM, Yu X, Anderson BP, Bradley A (2019) Snell's law for a vortex dipole in a Bose-Einstein condensate

50. Lee JG, McIlvain BJ, Lobb CJ, Hill WTI (2013) Analogs of basic electronic circuit elements in a free-space atom chip. Sci Rep 3:1034
51. Eckel S et al (2016) Contact resistance and phase slips in mesoscopic superfluid-atom transport. Phys Rev A 93:063619
52. Karpiuk T, Gremaud B, Miniatura C, Gajda M (2012) Superfluid fountain effect in a Bose-Einstein condensate. Phys Rev A 86:033619
53. Wilson KE, Samson EC, Newman ZL, Neely TW, Anderson BP (2012) Experimental methods for generating two-dimensional quantum turbulence in Bose-Einstein condensates. Ann Rev Cold At Mol 1:261
54. Seaman BT, Kramer M, Anderson DZ, Holland MJ (2007) Atomtronics: ultracold-atom analogs of electronic devices. Phys Rev A 75:023615
55. Catani J et al (2009) Entropy exchange in a mixture of ultracold atoms. Phys Rev Lett 103:140401

Chapter 5
A Tuneable Atomtronic Oscillator

5.1 Introduction

Using the trapping techniques presented in Chap. 4, we study superfluid transport in a tunable atomtronic circuit. The circuit studied consists of two reservoirs connected by a channel of tunable length and width. By exciting low amplitude plasma oscillations, the dependence of the oscillation frequency in response to the channel parameters is investigated. It is shown that a modified atomtronic circuit model, originating from acoustic modes, and that includes stray reservoir inductances well-describes the behaviour of the circuit throughout a wide parameter regime. Additionally, the resistive regime of the dynamics for a broad range of channel parameters is discussed, showing that a near-Ohmic resistive relationship is maintained. These results point to a simple phase-slip model of resistivity being relevant in these regimes, in contrast to the superfluid resistance model introduced by Feynman.

5.2 Background

Lumped-abstraction models are at the heart of engineering and electronics as they allow one to model the relevant macroscopic properties of a system, while remaining relatively simple through ignoring enough of the underlying microscopic physics. Recent advances in the trapping of ultracold degenerate atom systems have led to the field of *atomtronics*, where such lumped circuit models have been suggested to predict the behaviour of superfluid flow and particle transport in ultracold atomic

Part of the work presented in this chapter has been published by Science in the following publication: **G. Gauthier**, Stuart S. Szigeti, Matthew T. Reeves, Mark Baker, Thomas A. Bell, Halina Rubinsztein-Dunlop, Matthew J. Davis, and Tyler W. Neely, Quantitative Acoustic Models for Superfluid Circuits, *Physical Review Letters* **123**, 260402 (2019).

© The Editor(s) (if applicable) and The Author(s), under exclusive license to Springer Nature Switzerland AG 2020

G. Guillaume, *Transport and Turbulence in Quasi-Uniform and Versatile Bose-Einstein Condensates*, Springer Theses,
https://doi.org/10.1007/978-3-030-54967-1_5

systems [1, 2]. In this context, studies have included the transport of superfluid through weak links [3–6], superfluid transport through Josephson junctions in attractive and repulsive regimes for both fermions and bosons [7–11], and superfluid transport through mesoscopic channels [12, 13]. However, despite this body of work, and the support for several lumped abstraction models [13, 14], there is not yet agreement on the correct model of superfluid transport through a constriction [2].

It should be emphasised that the atomtronic situation can be remarkably different from its electrical analogues. In particular, the particle reservoirs are typically finite, in contrast to charge sources in an electrical circuit, which can be considered undepleted. Atomtronic circuits are thus expected to be subject to size effects, which may complicate the classification of the atomtronic system into isolated circuit elements. For Bose-Einstein condensate (BEC) systems, currents correspond to charge-neutral superfluid flow, so the microscopic details of resistive dissipation and bulk circuit quantities such as capacitance and inductance are also quite different in their origin. This has led to modified expressions for these quantities, such as the capacitance and kinetic inductance introduced in Ref. [14]. The nature of resistive dissipation was first theorised by Feynman [15], where dissipation is manifested in the energy required to create pairs of superfluid vortices, removing kinetic energy from the superfluid flow. Although supported by superfluid helium experiments [16], the highly-compressible nature of atomic BECs has meant the applicability of Feynman's model in these systems has remained an open question. The suitability of lump-abstraction models similarly requires further investigation. For example, recent work on the problem has suggested that for quantitative modelling the circuit model be abandoned in favour of a numerical approach [2].

A previous work by S. Eckel et al. [13] looked at the mapping of the flow between two reservoirs through a mesoscopic channel and wherever these could be modelled as equivalent to a resistor, capacitor, inductor, and Josephson junction circuit model. They achieved this by starting the system far from equilibrium, with all the atoms in the same reservoir. They concluded that the circuit model is an appropriate approximation, and proposed that the Feynman resistive model describes the resistance of the channel, the kinetic inductance serves as the inductance of the system, and the interaction energy results in a capacitance.

We experimentally and numerically explore the atomtronic properties of a similar "dumbbell" circuit, consisting of two reservoirs of fixed size, connected by a channel of broadly tuneable width and length (Fig. 5.1). We find that the low-bias oscillations of the circuit are quantitatively modelled as acoustic waves and the energy contained within these acoustic waves results in the circuit model. The acoustic waves extend throughout the channel and reservoirs. This, in contrast to the previous approaches [13, 14], where the channel and reservoirs were considered as separate inductor and capacitor elements respectively, we find that accurate modelling requires considering the entire trapped superfluid when calculating both quantities. This leads to the introduction of a "contact-inductance" where the channel terminates into the reservoirs, and one needs to also consider the channel capacitance for accurate modeling. Although the balance between, for instance, the channel capacitance

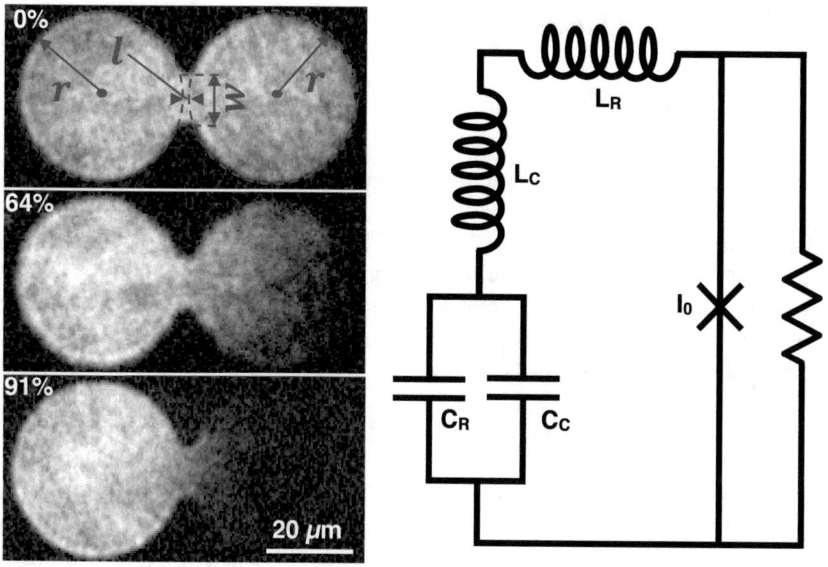

Fig. 5.1 Experimental setup and circuit. Left: in situ experimental density images of the atomtronic LRC circuit. The channel length l and channel width w are adjusted over a wide range of values, while the reservoir radius r is held fixed. By applying a magnetic field gradient, the system can be initialised with a chemical potential bias (lower panels). Right: Atomtronic circuit equivalent. The reservoirs and inductor both contribute capacitances (C_R, C_C) and inductances (L_R, L_C) respectively. In the low bias regime, where the current falls below the critical value I_0, the resistor is shunted by the Josephson junction. For large initial biases, creation of superfluid excitations leads to an Ohmic resistance R

and inductor capacitance, will change the relative contributions to the LC oscillation frequency as the reservoir size is increased, these stray contributions are not removed, and we find the reservoir-inductance to be a dominant factor, even for channels of moderate length. They thus cannot be attributed to size effects, but instead should be generally considered in the circuit analysis of any superfluid atomtronic system.

We also study the resistive damping of superfluid flow in the large initial-bias regime. We tune the channel width over a wide range, from $4\xi \rightarrow 100\xi$, where ξ is the superfluid healing length. Rather than the model of resistive shedding predicted by Feynman [15], we find a resistive relation close to Ohmic throughout the tuning range similar to Eckel et al. [13]. These results furthermore confirm that in this highly-excited regime the dissipation of superfluid flow is consistent with a simple phase-slip model [5, 13] which can lead to sound and vortex excitations.

5.3 Theoretical Models

5.3.1 Acoustic Model

In this section, we show low bias oscillations are described by acoustic sound-waves using the GPE.

Using Madelung transformation [Eq. (2.27)] of the GPE equation [Eq. (2.3)], one arrives at the hydrodynamic formulation given in Eq. (2.40a, b) (see Sect. 2.2.8 for details). Assuming small perturbations about a uniform hydrostatic equilibrium $V_{\text{ext}} = 0$, $n(\vec{r}, t) = n_0 + \delta n(\vec{r}, t)$ and $\vec{v} = \delta\vec{v}(\vec{r}, t)$, Eqs. (2.40a) and (2.40b) yield a wave equation for dispersive density waves

$$\partial_t^2 \delta n(\vec{r}, t) = (n_0 g/m^2)\vec{\nabla}^2 \delta n(\vec{r}, t) - (\hbar/2m)^2 \vec{\nabla}^4 \delta n(\vec{r}, t), \tag{5.1}$$

where higher order terms have been neglected. The second term may also be neglected due to small contribution in the long wavelength regime we are operating in, this yields an ordinary (non-dispersive) sound waves travelling at speed $c = \sqrt{n_0 g/m}$.

For small perturbations at long wavelengths, the system thus behaves as an ideal compressible fluid. An atomtronic circuit model can be obtained by developing a correspondence between the kinetic and potential energy stored within the sound waves and that of an LC circuit,

$$\frac{1}{2}(LI^2 + CV^2) = \frac{m}{2} \int \left(n_0 |\vec{v}|^2 + \frac{c^2 \delta n^2}{n_0} \right) d^2\vec{x}, \tag{5.2}$$

with the superfluid mass current assuming the role of I, and changes in pressure assuming the role of V. Assuming a symmetric step-density perturbation of the form

$$\delta n(\vec{r}) = \begin{cases} -\Delta N/V, & x < 0 \\ \Delta N/V, & x > 0 \end{cases} \tag{5.3}$$

with V being the total volume of the system. Then using Eq. (5.2), the capacitance between the two sides of the system and inductance are as in [13, 14] given by:

$$C = \frac{\Delta N}{\Delta \mu} \tag{5.4a}$$

$$L = \frac{m}{2I^2} \int n_0 |\vec{v}|^2 d^2\vec{x}. \tag{5.4b}$$

Since Eq. (5.2) is generally valid in this limit, a capacitance related to more general density perturbations could be derived.

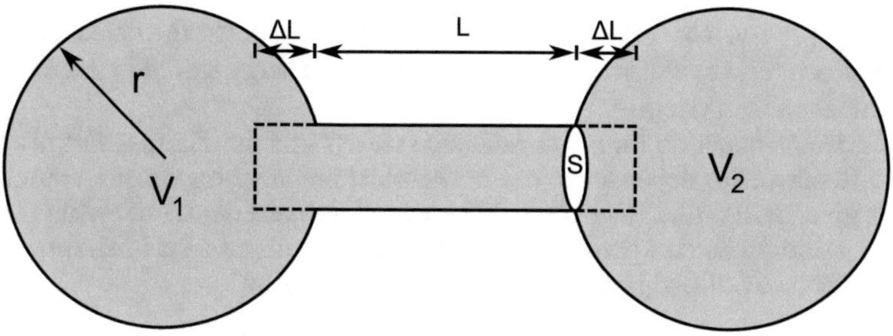

Fig. 5.2 Schematics of a modified two-spring Helmholtz resonator system connecting two reservoir through a channel. S is the cross-sectional area of the link, L is the length of the link, ΔL is the extra effective length added to the link to account for the kinematic inductance of the channel, V is the volume of the reservoir, and r is the radius of the reservoir

5.3.2 Acoustic Lumped Elements

Provided the wavelength of the sound is much larger than the characteristic scales of the device, all acoustic variables are constant over the dimensions of the device, and we may apply a lumped elements description as in the theory of acoustic circuits. The coupled Helmholtz resonator (Fig. 5.2) frequency is then given by

$$f = \frac{c}{2\pi}\sqrt{\frac{S}{L+\Delta L}\left(\frac{1}{\mathrm{V}_1}+\frac{1}{\mathrm{V}_2}\right)} \tag{5.5}$$

where $\mathrm{V} = 2\pi r^2 z_{\mathrm{TF}}$ is the volume of the reservoir, $S = 2wz_{\mathrm{TF}}$ is the cross sectional area of the link, and $\Delta l \propto \sqrt{S}$ to account for the extra flow outside of the channel and inside the reservoir. Note that while we have a flattened system, the lumped elements model is explicitly 3D. The behaviour of sound waves in this system have a 3D character; this is in contrast to the behaviour of vortices in the system that can be considered two-dimensional (see Chap. 6)

5.3.3 Chemical Capacitance

The capacitance of a standard electrical circuit describes the potential difference between two electrical reservoirs (one stores positive and the other negative charges) for a given difference in charge carrier between the two plates. From the Hamiltonian point of view it tracks the energy associated with charge imbalance in different parts of the system. In the case of a superfluid atomtronic circuit, a density imbalance between different parts of the system leads to a difference in the interaction energy stored per particle. The capacitance of an atomtronic circuit keeps track of how the difference in the number of atoms between different parts of the system affect the energy per particle. From the Hamiltonian point of view the capacitors in and atomtronic circuit should keep track of the potential energy stored in interactions of the particles in the system.

To derive an equation for the capacitance we assume the Thomas-Fermi approximation to derive the dependence of the chemical potential on the atom number the result is given in Eq. (2.18) where A is the area of one reservoir. Following [13], one can now write down an approximation for $\Delta\mu$ assuming a small increment of the atom number ΔN allowing use of the binomial expansion:

$$\Delta\mu = \beta\left[(N/2 + \Delta N/2)^{2/3} - (N/2 - \Delta N/2)^{2/3}\right] \approx (\beta N^{2/3})\frac{2^{4/3}\Delta N}{3N} = \frac{2^{4/3}\mu\Delta N}{3N}, \tag{5.6}$$

where β represents the constant prefactors in Eq. (2.18). Combining this with Eq. (5.2) leads to an expression for the capacitance:

$$C = \frac{\Delta N}{\Delta\mu} \approx \frac{3N}{2^{4/3}\mu}. \tag{5.7}$$

5.3.4 Resistive Models

The LC approximation is in general strictly valid in the low initial bias regime which is defined by the absence of energy dissipation from the oscillation. For larger initial biases, the fluid flow can exceed the local speed of sound of the superfluid, resulting in the resistive shedding of vortices [13], and higher frequency sound excitations. Above the critical current I_0, shedding of vortices and other excitations introduce a resistive element to the circuit model (Fig. 5.1). Feynman [15] first suggested this process for the case of infinite incompressible reservoirs with uniform density, where the excitation of pairs of quantised vortices lead to the dissipation of power [13, 15]:

$$P = \frac{\kappa\hbar I^2}{2w^2 n_{2D}}\ln\left(\frac{w}{\xi}\right), \tag{5.8}$$

where $\xi = \sqrt{\hbar/\mu m}$ is the healing length, κ a scaling constant to account for the finite extent of the system and image vortices, and I defines a uniform current through the channel. The conductance of the channel can be identified through $G_F = I^2/P$, and increases rapidly with increasing channel width being proportional to w^2. This is in contrast to a classical resistor, where $G \propto w$.

An alternative phase-slip model was introduced in Ref. [5], and supported by additional experiments [13]. Here, one considers that phase-slips associated with current $I > I_0$ result in the shedding of excitations, where the number of atoms involved in each excitation is $N_{ex} = w n_{2D}\xi = n_{1D}\xi$; we take ξ to be the average healing length over the channel. The expression for the conductance in this model is:

$$G_{PS} = C_0\frac{4\pi N_{ex}}{h\Delta\theta_c} = C_0\frac{4wn_{2D}\xi}{h}, \tag{5.9}$$

for a typical critical phase slip of $\theta_c \sim \pi$, where C_0 is a scaling constant to account for the finite system and the nature of the excitation. This expression thus recovers $G \propto w$ similar to a classical resistor.

5.4 Experimental Results

5.4.1 Experimental Setup

Our experimental setup, previously described in detail in Chap. 3, produces ^{87}Rb BEC in the $|\mathrm{F}=1, m_\mathrm{F}=-1\rangle$ state. The BEC is confined by a red-detuned optical sheet beam, with trapping frequencies $(\omega_z, \omega_r)/2\pi = (300, 6)$ Hz, and is vertically levitated by applying a magnetic field gradient of 30.5 G cm^{-1}, resulting in a planar superfluid. Configurable horizontal confinement is provided by 532 nm blue-detuned light, patterned using the Digital Micromirror Device (DMD) [17]. The DMD is utilized to create a hard-walled dumbbell-shaped potential. In-situ images of the BEC in the combined potential are shown in Fig. 5.1.

During the evaporation sequence a linear magnetic gradient is applied along the long axis of the dumbbell, creating an initial chemical-potential bias $\Delta\mu$. Due to the limited depth of the DMD potential, the atom number varies with the applied bias from $N \sim 2 \times 10^6$ for minimal biases, while a 100% bias results in $N \sim 5.5 \times 10^5$. The bias is suddenly turned off at time $t = 0$ ms. After some evolution time, a destructive, darkground-Faraday image (see Sect. 2.5.3) is taken to measure the reservoir population imbalance, $\eta_\mathrm{frac} = (N_\mathrm{L} - N_\mathrm{R})/(N_\mathrm{L} + N_\mathrm{R})$, calculated by dividing the image into two halves about the midpoint. Varying bias conditions are shown in Fig. 5.1. Here, N_L (N_R) refers to the number of atoms in the left (right) dumbbell reservoir.

5.4.2 Results: Undamped Plasma Oscillations

We first examine the behaviour of the circuit in the low initial-bias regime. The schematic circuit is shown in Fig. 5.1. In this regime, where the current is below the critical value I_0, underdamped Josephson plasma oscillations are expected [13]. The channel width w and length l are varied through a wide range, while fixing the reservoir radius $r = 20$ µm, and utilizing a small initial magnetic bias. On relaxing the initial bias at $t = 0$, repeatable sinusoidal oscillations result. The oscillation frequency was measured at several different atom numbers for each channel width and length and used this data to extract the frequency of the oscillation at a given atom number for each system parameters. A power law $\nu(N) = a_0 N^b + \nu_0$ was fit to the data (see Sect. 5.6.1); the experimental results in Fig. 5.3 result from using $N = 2 \times 10^6$ in the fitted function, and are shown along with the results of numerically

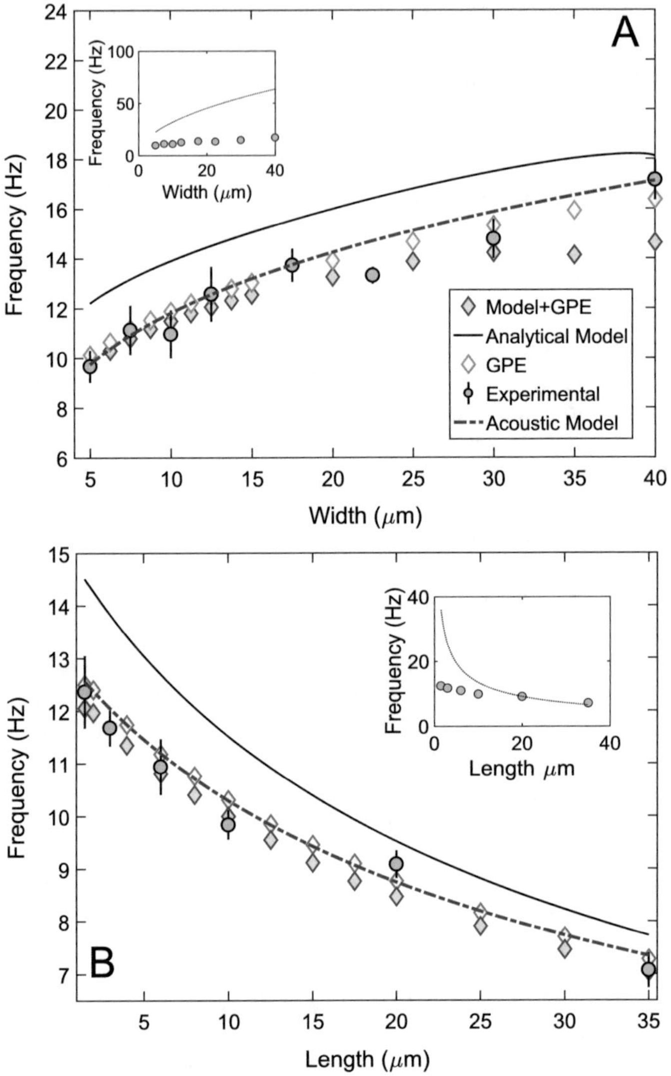

Fig. 5.3 Small amplitude frequency dependence. Oscillation frequencies for $\sim 2 \times 10^6$ atoms starting with ~2.5% initial bias as a function of **A** channel width for fixed 1.5 μm length and **B** channel length for fixed 12.5 μm width. The grey circles mark the experimental data, while the orange diamonds indicate the numerical results. The solid black lines represents the analytical model. Blue diamonds represent the analytical model, but conditioned with $\vec{u}$ and $\vec{I_c}$ from the GPE results. Purple dash-dotlines indicate the data fit to Eq. (5.5). Insets show the analytical result when the capacitance of the channel and contact inductance of the reservoirs is ignored [13, 14] (dotted lines), demonstrating the importance of considering the contact inductance

modelling Eq. (2.3) for the system. The experimentally observed frequencies are in excellent agreement with the GPE results performed by Dr Stuart Szigeti. We then fit the length Fig. 5.3B to Eq. (5.5) using $\Delta l = \sqrt{2 z_{\mathrm{TF}w}}$ and leaving z_{TF} and c as free parameters. The fit gives $c = 2.34(2)$ mm/s, and $z_{\mathrm{TF}} = 3.92(8)$ μm where the expected values are 2.60(5) mm/s and 2.3(1) μm, respectively. The extracted model is then plotted against the width dependence, see Fig. 5.3A, we find that predicted the frequencies are within 5% of the GPE modelled frequencies and are within error bars of the measured frequencies.

Assuming the atomtronic model with oscillation frequency $\nu = 1/\sqrt{LC}$, one can associate inductance $L = L_{\mathrm{C}} + L_{\mathrm{R}}$ and capacitance $C = C_c + C_{\mathrm{R}}$ defined by the system geometry and atom number. In Sect. 5.5, we show that the these quantities can be analytically modelled from Eq. (5.2), shown as solid lines in Fig. 5.3A, B. Accurate modelling requires the additional "contact inductance" L_R, associated with the fluid flow present in the reservoirs. This contribution is especially important for the short channels investigated here; the insets shown in Fig. 5.3 show that only including a channel inductance results in a overestimation of the frequency by a factor as much $\sim$2.8.

5.4.3 Results: Initial Currents Exceeding I_0

By increasing the initial bias, we can transition above the critical current I_0. With this initial condition, resistive shedding of excitations results [2, 13], and we image the resulting vortices with a 3 ms time-of-flight (TOF), as shown in Fig. 5.4B. Although there is significant vortex decay through annihilations, the resulting initial turbulent state is quite long lived, with several vortices remaining after 1000 ms. Despite the presence of the vortices and associated compressible (sound) excitations, we still observe repeatable sinusoidal oscillations, following the initial decay in η_{frac}. The oscillation amplitude initially follows the applied bias, while for slightly larger biases a maximum is reached as excitations begin to be shed. Similar amplitude oscillations are observed for very large biases, but we find reduced amplitudes for intermediate bias-values. A single-trajectory GPE simulation shows disordered behaviour for these parameters, which we attribute to higher-order excitation of the Bogoliubov (sound) modes of the system. Since the experimental data averages three experimental runs for each data point, this disorder in the flow is observed as reduced amplitude oscillations.

5.4.4 Resistive Regime

In this regime, the high degree of control over the channel parameters allows exploration of a wide range of channel widths. A 0.75 μm–15 μm range was utilized for a fixed channel length of 10 μm, with a 100% initial bias. In this regime the linear

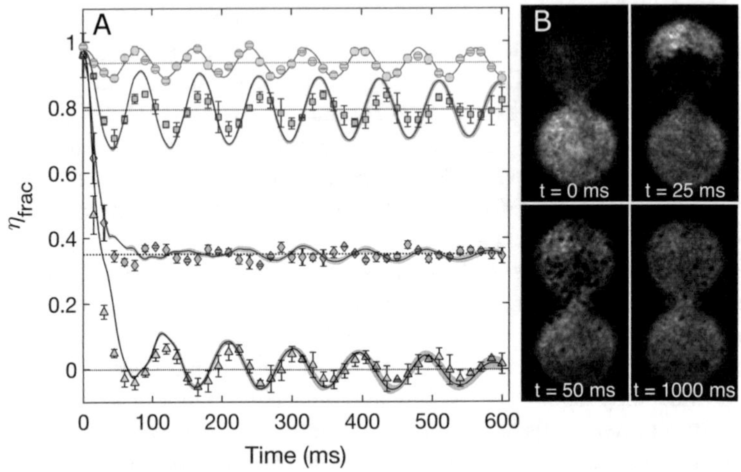

Fig. 5.4 Varying the initial bias. **A** Population imbalance evolution for a channel width of 12.5 μm and a length of 1.5 μm starting with different initial biases; circles (2.5%), squares (18%), diamonds (60%), triangles (96%). The solid lines are numerical simulations performed by Dr. Stuart Szigeti, specifically a single GPE simulation of atom number $N = 2 \times 10^6$ (2.5%) and the average of ten GPE simulations with initial N stochastically chosen from a Gaussian distribution of mean $N = 1.6 \times 10^6$ (18%), 1.4×10^6 (60%), 1.06×10^6 (96%) and variance $(N/10)^2$ (standard errors indicated by shading). This stochasticity accounts for the ~10% shot-to-shot atom number fluctuations observed in the experiment. The data is offset for clarity, the equilibrium condition corresponds to $\eta_{\rm frac} = 0$. Both the 60% and 96% cases required a slight negative bias to attain a $\eta_{\rm frac} = 0$ equilibrium; without this, $\eta(t)$ oscillates around a nonzero offset due to an imbalance of shed vortices caused by the higher chemical potential for the same number of atoms in the reservoir with more vortices. **B** 3 ms time of flight (TOF) images of the reservoir atomic distribution after hold times indicated with an initial bias of 63%; vortices can be clearly seen as dark dips in the Faraday images of the BECs

analysis is not expected to hold, with the critical current I_0 exceeded, leading to the shedding of excitations as seen in Fig. 5.4.

We extract the conductance of the channel experimentally and numerically by fitting the decay in $\eta_{\rm frac}$ (see Sect. 5.6.1), with an RC decay model. The conductance $G = 1/R$ and corresponding $n_{\rm 1D}$ is calculated (see Sect. 5.6.2) for each channel width, shown in Fig. 5.5A. Since the GPE data allows quantitative extraction of the healing lengths, and thus $N_{\rm ex} = n_{\rm 1D}\xi$, we fit a linear function to the phase-slip model $G_{\rm fit} = G_{\rm PS} + G_0$, allowing determination of C_0, Fig. 5.5B. We find values of $C_0 = 4.4 \pm 0.1$ and $G_0 = -1.3 \pm 0.3 \times 10^{36}\ {\rm J^{-1}s^{-1}}$ result in an r-squared value for the fit of ~0.998. By contrast, a similar fit to the Feynman model yields $\kappa = 10 \pm 1$ and an r-squared value of ~0.962; the resulting fits are shown in Fig. 5.5B and residuals in Fig. 5.5C.

These results suggest that the phase-slip model better describes the dynamics of the system. This may be due to the importance of compressible dissipative channels for the BEC system. For example, the smallest (most resistive) channels explored show

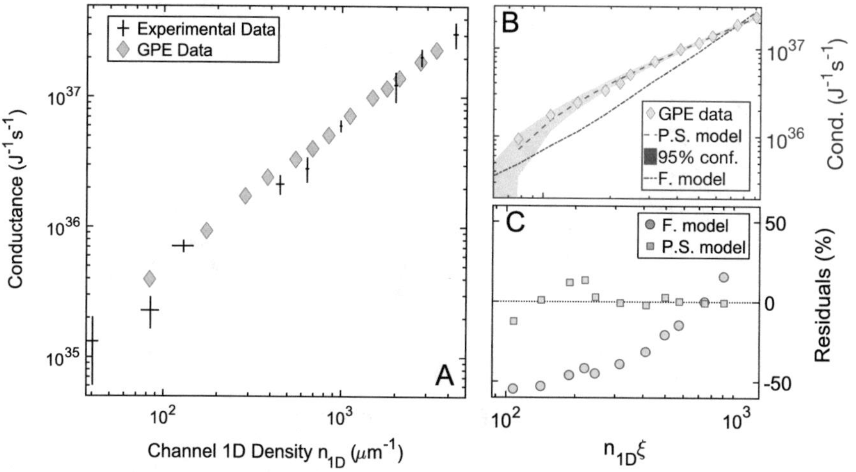

Fig. 5.5 Channel conductance. **A** The channel conductance is shown versus n_{1D}, for the GPE simulations and experiment, with excellent agreement. **B** GPE conductance vs. $n_{1D}\xi = wn_{2D}\xi$, fit with the phase-slip model, resulting $C_0 = 4.4 \pm 0.1$; the fit residuals are shown in **C**. By contrast, the Feynman conductance model best-fit exhibits significant residuals

little to no vortex shedding throughout the dynamics. The non-zero and negative G_0 suggests a threshold n_{1D} is required for conduction; this may be due to the transverse confinement of the channel approaching the chemical potential μ. In the experiment, the minimum conductance, as in Fig. 5.5A, appeared limited to non-zero values from distillation of the condensate between the reservoirs, present even in the absence of a channel, with transport facilitated by the presence of the thermal cloud [18].

5.4.5 Thermal Damping

We return to the low bias regime and examine the dependence of the thermal damping of low-amplitude oscillations on the BEC fraction, $N_{\rm cond}/N$. We achieve lower BEC fractions by ending the final evaporative ramp of the optical dipole trap at a higher trap depth, before relaxing quickly to the final trap value during the step where the cloud is levitated against gravity; we find that this final step occurs over time-scales too short for efficient evaporation, leading to similar total atom number in the final trap, but with a smaller BEC fraction (higher temperature) with similar $N_{\rm cond}$. An increased thermal fraction is expected to increase particle exchange between the coherent condensate and incoherent thermal cloud, resulting in redistribution of particles leading to equilibration of the initial chemical potential offset. We indeed find increased damping of the oscillations with decreasing condensate fraction, and fit an exponentially decaying envelope; an example oscillation at the lowest condensate fraction is shown in Fig. 5.6A, while the decay constant variation with BEC fraction is shown in Fig. 5.6B.

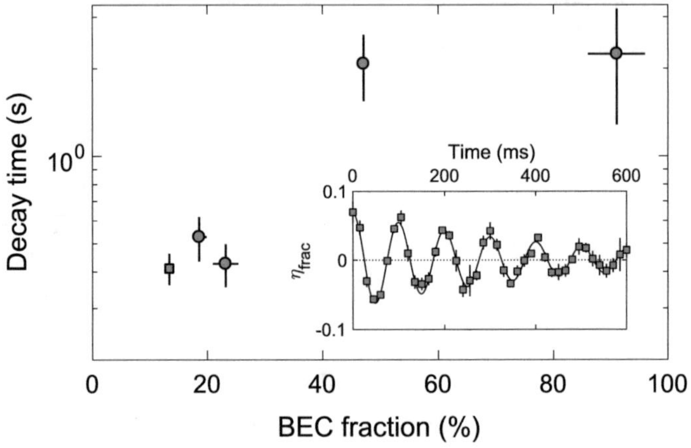

Fig. 5.6 Thermal damping. The BEC fraction is reduced by arresting the final evaporative ramp to higher final values. For low BEC fractions, the oscillation damping increases, resulting in a decreased decay time (inset: the lowest BEC fraction of 13% represented by a square data point). Error bars indicate the 95% confidence intervals of the fit, for slow decay, the limited data time-series leads to the large error bars

5.5 Modelling the Dumbbell Inductance

In this section, we describe the analytical model leading to the solid lines in Fig. 5.3. This is done by modelling the capacitance using Eq. (5.7), and modelling the superfluid current in the system to obtain an accurate measure of the kinetic inductance.

The inductance is used to keep track of the kinetic energy of the system as a function of the atom transport rate through the channel. One of the main assumptions is that the current in the system scales linearly with the current through the channel. The macroscopic measure of interest is the rate of transport of the atoms through the channel (I_c) so we will develop the inductance of the channel which includes a kinetic contact inductance due to the flow in the reservoir to fill up the channel and the flow of atoms in the channel itself. Looking at the current contours from GPE simulations of the transport in our reservoir system, see Fig. 5.7B, we find three distinct regions. The first region, A shows spherical contours connected to the reservoir edges, while B shows oval contours linked to the reservoir edges, and C shows a semi constant velocity going through the channel.

To derive the current in our system we assume three different currents one for each of the constant current contours of the form:

$$I(x) = \begin{cases} I_{\text{Planar}}(x), & 0 \leq x \leq l' \\ I_{\text{Oval}}(x), & l' \leq x \leq l' + d/2 \\ I_{\text{Circular}}(x), & l' + d/2 \leq x \leq l' + d/2 + r \\ -I(-x), & x < 0 \end{cases}, \tag{5.10}$$

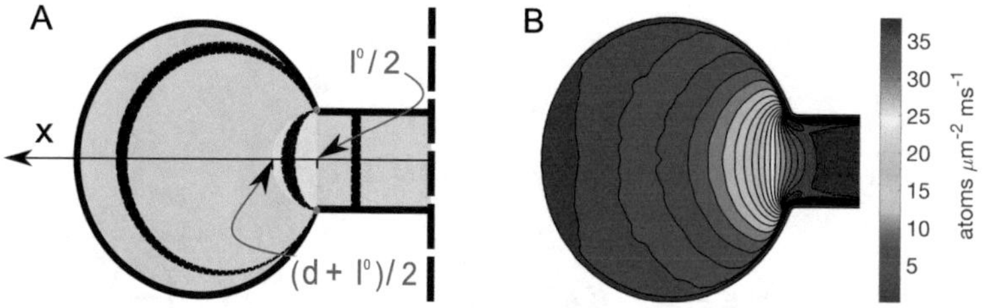

Fig. 5.7 Surface of constant current reservoir transport. **A** Depiction of the different surface elements used to approximate the channel and reservoir currents. The red-shaded region uses spherical constant-current wavefronts (i.e. $x > (d + l')/2$), the green shaded region uses parabolic-constant current wavefronts (i.e. $l' \leq x \leq l' + d/2$), while the blue shaded region uses planar wavefronts. **B** For comparison, the 2D current density at $z = 0$ is shown, as determined from the GPE simulations. The data are averaged over a 10-ms window, centered around the time of peak superfluid velocity through the midpoint of the channel

where l' is the effective length of the channel given by:

$$l' = l - r\left(\sqrt{4 - \frac{d^2}{r^2}} - 2\right). \tag{5.11}$$

Using the continuity equation, Eq. (2.40b), we know that the rate at which the number of atoms inside a volume is changing is given by the difference between the current entering the volume and exiting the volume which means that the rate at which the density changes inside the volume is given by the derivative of the current across the surface divided by the length of the surface interface

$$\frac{dn_{2D}(x)}{dt}\frac{dA(x)}{dx} = \frac{dI(x)}{dx} \tag{5.12}$$

where dA(x) is the constant current surface area element depicted in Fig. 5.7 by the dark area for the three different regions of the reservoir and has the functional form:

$$dA(x) = \begin{cases} d \cdot dx, & 0 \leq x \leq l' \\ \frac{d\pi}{4} dx, & l' \leq x \leq l' + d/2 \\ 2(x - l')\left[\pi + \frac{d}{\sqrt{4r^2 - d^2}} - \arcsin\left(\frac{d}{2(x - l')}\right)\right] dx, & l' + d/2 \leq x \leq l' + d/2 + r \\ dA(-x), & x < 0 \end{cases}. \tag{5.13}$$

We assume that the rate of change of the density is opposite and symmetric about the center of the reservoir, and that the change in the density is uniform and fully contributes to the current going being transported from one reservoir to the other:

$$\frac{dn_{2D}(x)}{dt} = \begin{cases} 2I_c/A, x \le 0 \\ -2I_c/A, x \ge 0 \end{cases}, \tag{5.14}$$

where A is the area of the dumbbell system given in by:

$$A = 2r^2\left(\pi - \arcsin\frac{d}{2r}\right) + d*l + 2*r*d\left(1 - \frac{1}{4}\sqrt{4 - \left(\frac{d}{r}\right)^2}\right), \tag{5.15}$$

and $I_c = d\Delta N/dt$ is rate of change in the number particles between the reservoir as a function of time.

A discontinuity in the current would mean an infinite rate of change in the derivative of the density, which is not physically allowed, implying that the current must be continuous everywhere, giving

$$I(x^-) = I(x^+). \tag{5.16}$$

From Eqs. (5.12) and (5.14), we also get that the product of the derivative and the inverse area elements should be a constant and continuous for $x > 0$ and $x < 0$ with a discontinuity at $x = 0$:

$$\left.\frac{dI(x)}{dx}\left[\frac{dA(x)}{dx}\right]^{-1}\right|_{x\to x^-} = \left.\frac{dI(x)}{dx}\left[\frac{dA(x)}{dx}\right]^{-1}\right|_{x\to x^+} \forall x \neq 0. \tag{5.17}$$

Since there can be no flow across the boundary, the current go to zero at the boundary giving the condition:

$$I_{\text{Boundary}} = I(l' + d/2 + r) = I(-(l' + d/2 + r)) = 0. \tag{5.18}$$

By mass conservation argument, the current in the center of the channel must be equal to the rate of change of the difference in the number of atoms between the two reservoirs (i.e. $I(0) = I_c = d\Delta N/dt$). Using these conditions, we can derive an analytical formula for the current in the reservoirs as a function of I_c.

The effective kinetic inductance for our system can be obtained by equating the energy contained in the inductor to the full kinetic energy of the system as derived in the first half of Eq. (5.2):

$$\frac{1}{2}LI_c^2(t) = \frac{1}{2}\int_x |\vec{v}(\vec{r}, t)|^2\, dm \tag{5.19}$$

Defining a surface ($C(x)$) as surfaces of constant current:

$$C(x) = \begin{cases} d, & 0 \leq x \leq l' \\ \pi\sqrt{\frac{(x-l')^2+(d/2)^2}{2}}, & l' \leq x \leq l' + d/2 \\ 2(x-l')(\pi - \arcsin\left[\frac{d}{2(x-l')}\right]), & l' + d/2 \leq x \leq l' + d/2 + r \\ A(-x), & x < 0 \end{cases}, \tag{5.20}$$

and assuming that the current is constant over the surfaces, that there is only flow perpendicular to the surface contours and that the density is uniform throughout the system. Then the current can be expressed in terms of the velocity as

$$I(s) = n_{2\text{D}} C(x)|\vec{v}(x)|. \tag{5.21}$$

With the assumption of constant density, the mass element (dm) in Eq. (5.19) can also be expressed in terms of density and constant current surface area element Eq. (5.13) as

$$dm = n_{2\text{D}} dA(x). \tag{5.22}$$

Combining Eq. (5.19) with Eqs. (5.21) and (5.22) we can isolate an analytical formula for the inductance in the system in terms of system geometry only:

$$L = n_{2\text{D}} \int_{-(l'+d/2+r)}^{l'+d/2+r} \frac{I^2(x)}{I_c^2 C^2(x)} dA(x) \tag{5.23}$$

where m is the mass of a single atom, $C(x)$ was defined in Eq. (5.20), and $n_{2\text{D}}(s) = N/A$ is the 2D density which is assumed to be equal over the whole reservoir.

The inductance of the system can be calculated from the GPE numerics using a time averaged Eq. (5.19) where the right side is integrated numerically over a full oscillation period (T) and I_c is given by

$$I_c(t) = \frac{1}{T} \int_0^T \iint dydz n_{2\text{D}} \left|\vec{v}(\vec{r}, t|x=0) \cdot \hat{x}\right|^2 dt. \tag{5.24}$$

The time-averaging is done to prevent residual kinetic energy at low current from biasing the measure of the inductance. The GPE calculated kinetic inductance multiplied by the analytical capacitance [Eq. (5.7)] are plotted in Fig. 5.3 a blue diamonds.

5.6 Extracting the System Parameters

5.6.1 Fitting Function for Decay Constant

For the GPE numerical simulations, the decay constant τ was extracted by fitting the function $y = (1 - A)\exp(-t/\tau) + B\cos(2\pi\omega_0 t + E) + A$ to $\eta_{\rm frac}(t)$. This assumes that at higher biases, the undamped LC oscillations are modified by an initial exponential decay associated with capacitative discharge. A similar fitting procedure was considered in Ref. [2], which included functions that smoothly turned the capacitive discharge and LC oscillation off and on, respectively; this was specifically of the form: $y = [1 + \exp(\frac{t-t_c}{t_w})]^{-1}\exp(-t/\tau) + [1 + \exp(-\frac{t-t_c}{t_w})]^{-1}[B\cos(2\pi\omega_0 t + E) + A]$. We compared the τ obtained using both fitting functions and found broad agreement within statistical uncertainty. We therefore opted to use the former fit, as it contained fewer parameters.

For the experimental data, the decay constant τ was extracted by fitting the function $y = (1 - A)\exp(-t/\tau) + A$ to $\eta_{\rm frac}(t)$, the absence of the term $B\cos(2\pi\omega_0 t + E)$ is due to the absence of observed oscillation after equilibration.

For the two smallest channel widths considered, 1.5 μm and 2.0 μm, a simple exponential fit was used to extract the decay constant, since in both these cases, the decay was sufficiently slow that simulations over time intervals of 5–10 s only provided data for this initial decay. Running these simulations for long enough time intervals to obtain data for the LC oscillations was not computationally viable.

5.6.2 Averaged 1D Density of Channel

From the GPE simulations, we instantaneous 1D density is given by $n_{\rm 1D}(x,t) = \int dydz|\psi(\vec{r},t)|^2$. However, when analyzing the conductance we are interested in the 1D density *within the channel*. We estimate this quantity by averaging $n_{\rm 1D}(x,t)$ over the bulk of the channel; for the 10 μm channel length considered here, we averaged from $x = -4\,\mu$m to $x = +4\,\mu$m: $n_{\rm 1D}(t) = \frac{1}{8\mu{\rm m}}\int_{-4\,\mu{\rm m}}^{+4\,\mu{\rm m}} n_{\rm 1D}(x,t)$. By taking the time average of this quantity from the time point at the first turning point of $\eta_{\rm frac}$, which is just after the large initial decay due to capacitative discharge, we obtain the time averaged 1D channel density, which forms the horizontal axis for the simulation data in Fig. 5.5. These results were compared to extracting the density from the healing length at the end of the channel and both methods were in agreement. GPE and experimental channel density data for the simulated and measured channels is shown in Appendix A, in Tables A.3 and A.4, respectively.

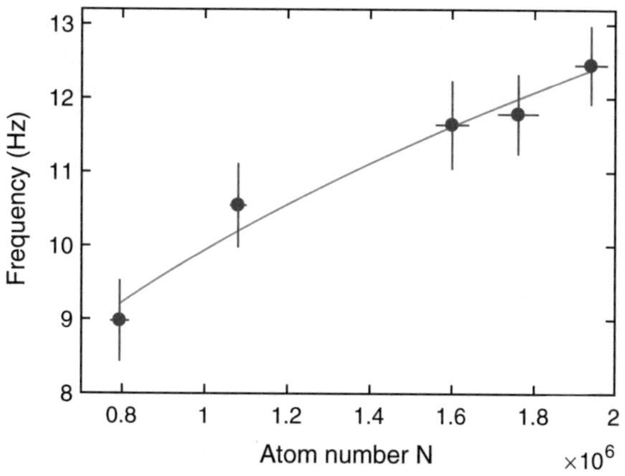

Fig. 5.8 Frequency variation with atom number for the $w = 12.5$ μm and $l = 1.5$ μm channel case. The solid line is a power law fit to the data; see text

5.6.3 Frequency Variation with Atom Number

In the low bias regime, the oscillation frequency $f = 1/2\pi\sqrt{LC}$ varies with atom number through Eqs. (5.7) and (5.23), leading to $\nu \propto N^{-1/3}$. The atom number contained within the dumbbell furthermore varies with the channel width and length, as the enclosed area changes, resulting in different initial loading conditions. In order to consistently compare oscillation frequencies for the different channel widths and lengths shown in Fig. 5.3, multiple data sequences at each condition were taken, where the atom number was varied. The frequency dependence was then extracted through a power-law fit $\nu(N) = a_0 N^b + \nu_0$. Figure 5.8 shows examples of the experimental data and fit, for the $w = 12.5$ μm and $l = 1.5$ μm channel, resulting in $b = 0.3 \pm 0.1$. In Fig. 5.3, the frequencies were determined by inputting $N = 2 \times 10^6$ into these fitted functions; the error bars shown on the plot represent the 95% confidence intervals from the fit. The full set of data corresponding to the different widths and lengths, for varying atom numbers, is shown in Appendix A, Tables A.1 and A.2.

5.7 Conclusion

In summary, we investigated the full dynamics of a dumbbell atomtronic circuit, demonstrate their similarity to an acoustic Helmholtz resonator, and how they can be reduced to a lumped circuit element model. We find that modelling the inductance of the system requires careful consideration of the full flow field, although an analytic approach can be developed to assign the inductance for particular circuit elements. Above the critical current, we find an Ohmic conductive relationship across the entire parameter regime, suggesting that observing Feynman-like resistance will require a situation closer to the ideal model, such as remaining in the incompressible limit by maintaining quasi-uniform superfluid density. Our results will be useful for the potential applications of atomtronic circuits, such as inertial sensors, and are furthermore important for applications utilising other superfluid systems, such as circuits of superfluid helium.

References

1. Chien C-C, Peotta S, Di Ventra M (2015) Quantum transport in ultracold atoms. Nat Phys 11:998–1004
2. Li A et al (2016) Superfluid transport dynamics in a capacitive atomtronic circuit. Phys Rev A 94:023626
3. Moulder S, Beattie S, Smith RP, Tammuz N, Hadzibabic Z (2012) Quantized supercurrent decay in an annular Bose-Einstein condensate. Phys Rev A 86:013629
4. Wright KC, Blakestad RB, Lobb CJ, Phillips WD, Campbell GK (2013) Driving phase slips in a superfluid atom circuit with a rotating weak link. Phys Rev Lett 110:025302
5. Jendrzejewski F et al (2014) Resistive flow in a weakly interacting Bose-Einstein condensate. Phys Rev Lett 113:045305
6. Eckel S, Jendrzejewski F, Kumar A, Lobb CJ, Campbell GK (2014) Interferometric measurement of the current-phase relationship of a superfluid weak link. Phys Rev X 4:031052
7. Albiez M et al (2005) Direct observation of tunneling and nonlinear self-trapping in a single bosonic Josephson junction. Phys Rev Lett 95:010402
8. Levy S, Lahoud E, Shomroni I, Steinhauer J (2007) The a.c. and d.c. Josephson effects in a Bose-Einstein condensate. Nature 449:579–583
9. Valtolina G et al (2015) Josephson effect in fermionic superfluids across the BEC-BCS crossover. Science 350:1505
10. Spagnolli G et al (2017) Crossing over from attractive to repulsive interactions in a tunneling bosonic Josephson junction. Phys Rev Lett 118:230403
11. Burchianti A et al (2017) Connecting dissipation and phase slips in a Josephson junction between fermionic superfluids. ArXiv e-prints
12. Stadler D, Krinner S, Meineke J, Brantut J-P, Esslinger T (2012) Observing the drop of resistance in the flow of a superfluid Fermi gas. Nature 491:736–739
13. Eckel S et al (2016) Contact resistance and phase slips in mesoscopic superfluid-atom transport. Phys Rev A 93:063619
14. Lee JG, McIlvain BJ, Lobb CJ, Hill WTI (2013) Analogs of basic electronic circuit elements in a free-space atom chip. Sci Rep 3:1034
15. Feynman RP (1955) In: Gorter CJ (ed) Elsevier, pp 17–53
16. Varoquaux E (2015) Anderson's considerations on the flow of superfluid helium: some offshoots. Rev Mod Phys 87:803
17. Gauthier G et al (2016) Direct imaging of a digital-micromirror device for configurable microscopic optical potentials. Optica 3:1136–1143
18. Shin Y et al (2004) Distillation of Bose-Einstein condensates in a double-well potential. Phys Rev Lett 92:150401

Chapter 6
Creation and Dynamics of Onsager Vortex Clusters

6.1 Introduction

Adding energy to a system through transient stirring usually leads to more disorder. However, in a bounded two-dimensional fluid containing point-like vortices, Onsager found a surprising result: increasing its energy leads to highly-ordered, persistent vortex clusters. In this chapter, we demonstrate the realization of these vortex clusters in a planar superfluid. Despite their high energy, we demonstrate that they persist for long times, maintaining the superfluid system far from global equilibrium. Our experiments explore a regime of vortex matter at negative absolute temperature, opening new directions for research in two-dimensional turbulence, systems with long-range interactions, and the dynamics of topological defects. Our results have relevance to systems such as helium films, nonlinear optical materials, fermion superfluids, and quark-gluon plasmas.

6.2 Background

An isolated system that is initially stirred will generally undergo dynamics that eventually achieve quiescent thermodynamic equilibrium. However, in some systems, the near decoupling of particular degrees of freedom can lead to a separation of time-scales for equilibration [1]. This decoupling can result in strikingly different thermodynamic behaviour within sub-systems, where no spatially-uniform equilibria exist [2–4]. As first recognized by Lars Onsager [4], a prototypical example is a sys-

Part of the work presented in this chapter has been published by Science in the following publication: **G. Gauthier***, M. T. Reeves*, X. Yu, A. S. Bradley, M. Baker, T. A. Bell, H. Rubinsztein-Dunlop, M. J. Davis, T. W. Neely, Giant vortex clusters in a two-dimensional quantum fluid, *Science*, **364**, 1264–1266 (2019).

© The Editor(s) (if applicable) and The Author(s), under exclusive license to Springer Nature Switzerland AG 2020

G. Guillaume, *Transport and Turbulence in Quasi-Uniform and Versatile Bose-Einstein Condensates*, Springer Theses,
https://doi.org/10.1007/978-3-030-54967-1_6

tem of N point vortices [5] contained within a bounded two-dimensional (2D) fluid. Onsager's model predicts that with sufficient decoupling between two and three-dimensional flow, and negligible viscous dissipation, high energy incompressible flow leads to low-entropy, large-scale aggregations of like-circulation vortices. This is strikingly different from the behaviour of vortices in 3D fluids [6, 7]. Onsager's theory has been highly influential [2, 3], providing some understanding of diverse classical quasi-2D systems such as turbulent soap films [8], guiding-centre plasmas [9], self-gravitating systems [10], and Jupiter's Great Red Spot [11]. However, despite the intuition provided by Onsager's model, quantitative demonstrations of point-vortex statistical mechanics have been elusive. While two-dimensional classical fluid dynamics can realize vortex cluster growth, their vortices are continuous and cannot be realistically modelled by discrete points [12]. Aware of this limitation, Onsager noted the model would be more realistic for 2D superfluids, where vortices are discrete, with circulations constrained to $\Gamma = \pm h/m$, where h is Planck's constant and m is the mass of a superfluid particle. The physical manifestation of high-energy point-vortex clusters in any fluid system has however remained unrealized since Onsager's seminal work in 1949 [4].

The incompressible kinetic energy of an isolated 2D fluid containing N point vortices can be expressed in terms of the relative vortex positions [5]. In an unbounded uniform fluid, it has the form given in Eq. (2.48), Onsager's key insight was that since Eq. (2.48) is determined by the positions $\vec{r}_i$, for a *confined* fluid, the available phase space becomes bounded by the area of the container [4]. This property dramatically alters the system's thermodynamic behaviour.

The equilibrium phases of point-vortex matter in a bounded region are shown schematically in Fig. 6.1. Thermodynamic equilibria maximize the entropy $S(E)$, which describes the logarithm of possible configurations of vortices and antivortices at a given energy E. The vortex temperature is given by $T = (\partial S/\partial E)^{-1}$. The low energy, positive temperature phase ($T > 0$) consists of bound vortex-antivortex pairs (Fig. 6.1A). As the energy increases these pairs unbind [13], until the vortex distribution becomes completely disordered (Fig. 6.1B), marking the point of maximum entropy ($T = \infty$). However, due to the bounded phase space, this point occurs at finite energy; at still higher energies vortices reorder into same-sign clusters [2, 4], thus decreasing the entropy, and yielding negative absolute temperatures ($T < 0$). At sufficiently high energies the system undergoes a clustering transition ($T = T_c$) [14]; here the vortices begin to *polarize* into two giant clusters of same-circulation vortices (Fig. 6.1C), whose structures are determined by the shape of the container. In the limit $E \to \infty$ the clusters shrink to two separated points (Fig. 6.1D), and the temperature approaches the limiting supercondensation temperature ($T \to T_s$), which is independent of geometry [15]. For vortices in a superfluid, where the vortex core size is non-zero and determined by the healing length ξ (see Sect. 6.4.6), core-repulsion at lengths $\sim\xi$ prevents the vortex clusters from collapsing to points at infinite energy.

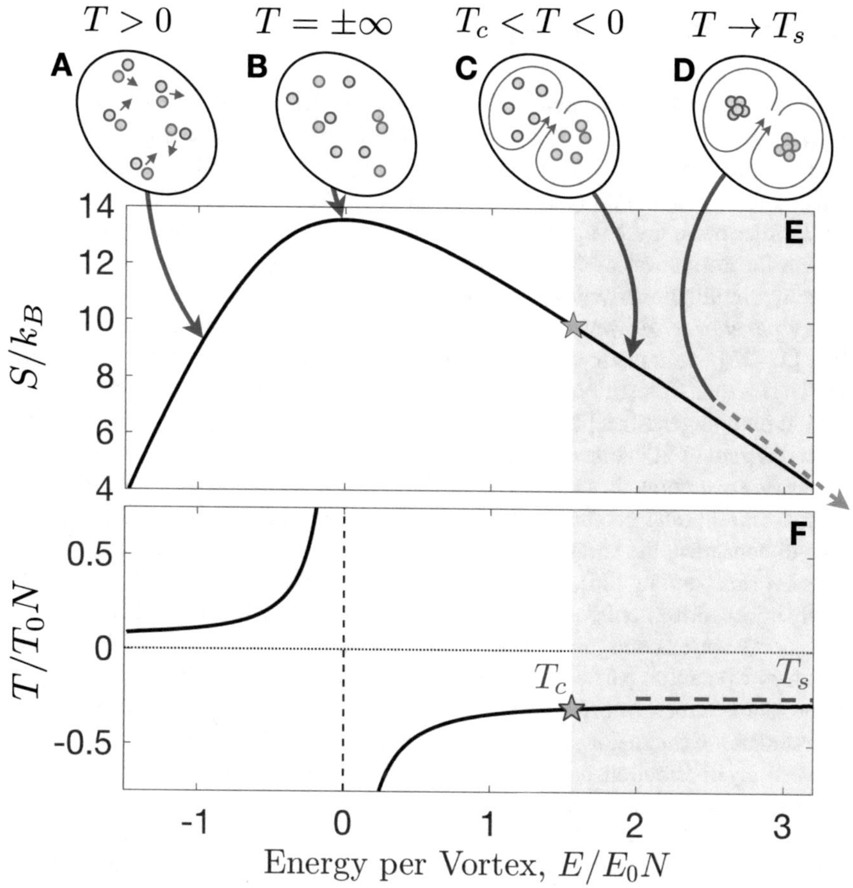

Fig. 6.1 Phases of point-vortex matter in a bounded domain. **A** Small positive temperatures exhibit bound vortex-antivortex pairs. **B** As the vortex temperature $T \to \infty$ vortex positions become uncorrelated. **C** At high enough energies a clustering temperature T_c is reached where giant Onsager vortex clusters form. **D** As $E \to \infty$ the clusters shrink to two separated points forming a supercondensate. **E, F** Entropy and temperature versus energy curves for a neutral vortex gas ($N_+ = N_- = N/2 = 9$), generated by Monte Carlo simulations for the elliptical domain shown (see text). The shaded region contains Onsager vortex clusters and the purple star marks the clustering transition temperature T_c calculated from mean-field theory (see Sect. 6.4.8). The red line indicates the supercondensation limit $E \to \infty$, $T \to T_s = -0.25T_0N$. The energy and temperature units are $E_0 = \rho_0\Gamma^2/4\pi$ and $T_0 = E_0/k_B$ respectively, where k_B is Boltzmann's constant

6.3 Experimental Results

To physically realize this idealized model, the vortices must form a well-isolated subsystem and effectively decouple from the other fluid degrees of freedom. A large and uniform 2D Bose-Einstein condensate (BEC) near zero temperature, with weak vortex-sound coupling, has been proposed as a suitable candidate system [16–19]. Furthermore, these superfluids allow for vortex-antivortex annihilation, which favours the formation of Onsager vortices through evaporative heating [18, 20], whereby annihilations remove low energy dipoles, thus increasing the remaining energy *per vortex*. However, while small transient clusters have been observed in BEC [21–23], attempts to create Onsager's vortex clusters have thus far been hindered by thermal dissipation and vortex losses at boundaries [24], which are enhanced by fluid inhomogeneities [25]. This has prevented the experimental study of the full phase diagram of 2D vortex matter shown in Fig. 6.1.

Here we overcome these issues by working with a uniform planar ^{87}Rb BEC confined to an elliptical geometry as described in Sect. 6.4.2. While the BEC itself is three-dimensional, the vortex dynamics are two-dimensional due to the large energy cost of vortex bending [26]. By engineering different stirring potentials, we can efficiently inject vortex configurations with minimal sound excitation (see Sect. 6.4.5). A high energy vortex configuration can be injected using a double-paddle stir, whereby two narrow potential barriers [27, 28] are swept along the edges of the trap. Due to the broken symmetry of the ellipse, the maximum entropy state is a vortex dipole separated along the major axis [29]. The stirring protocol is well mode-matched to this vorticity distribution, and we find the vortices rapidly organize into two Onsager vortex clusters (Fig. 6.2B).

We contrast these results with the injection of a low energy configuration from sweeping a grid of smaller circular barriers through the BEC. Experimentally we find this results in a similar number of vortices (Fig. 6.2D, E), but in a disordered distribution that is a candidate for evaporative heating [18, 25, 30] (c.f. Fig. 6.1B). Gross-Pitaevskii equation (GPE) simulations, performed by Dr. Matthew Reeves, quantitatively model both stirs and are compared in Fig. 6.2C, F.

While vortex sign detection [24, 30] is possible (see Sect. 6.5), the clustered states are *non-uniform* equilibria, and their presence can be confirmed from the (unsigned) vortex density $\rho = \sigma_+ + \sigma_-$, where σ_+ (σ_-) denotes the distribution of positive (negative) vortices (see Sect. 6.4.9). Figure 6.3A displays a time-averaged position histogram, generated by measuring the experimental vortex positions at one-second intervals over ten seconds of hold time following injection. As expected for our elliptical geometry (see Sect. 6.4.9), the density shows two distinct persistent clusters separated along the major axis. The clusters remain distinguishable up to 9 s of hold time in individual frames. By contrast, the grid stir in Fig. 6.3B shows a nearly uniform distribution of vortices consistent with an unclustered phase (Fig. 6.1A, B). Figure 6.3C, D show the corresponding (signed) density $\omega = \sigma_+ - \sigma_-$ from GPE simulations, showing polarized clusters for the paddle stir, contrasted with $\omega \approx 0$ for the grid stir. Figure 6.3E compares the total vortex number as a function of time for the

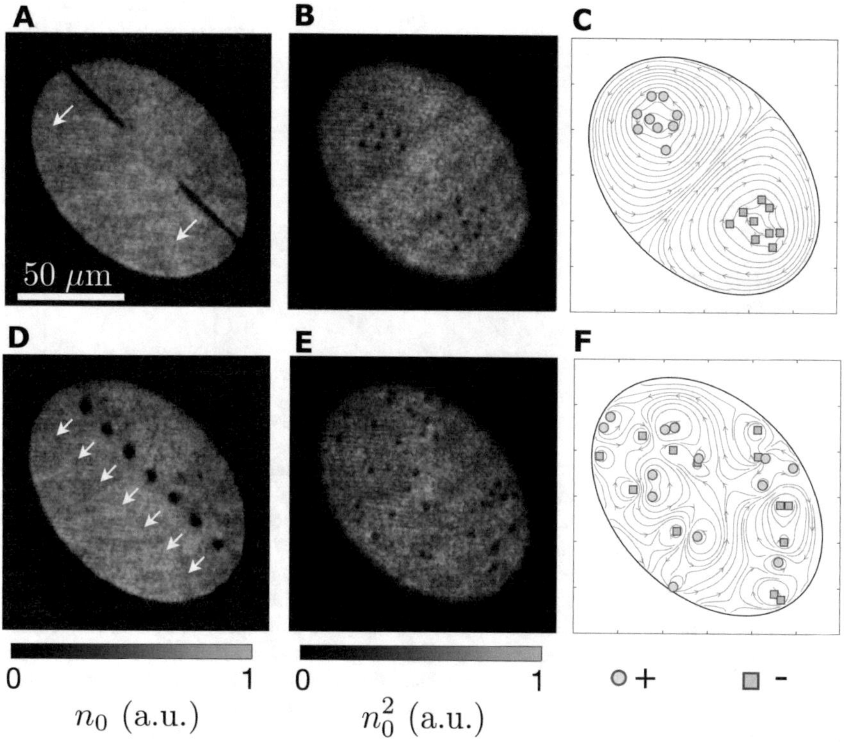

Fig. 6.2 Experimental vortex injection. **A** Two large paddle potentials stir the BEC inducing large-scale flow (In situ image, part-way through the stir). **B** A 3 ms time-of-flight Faraday image directly after the paddle stir clearly resolves injected vortices (see) localized into two clusters. **C** Simulation of the paddle stir showing velocity contours, with the location and circulations of the vortices demonstrating the injection of a clustered vortex dipole. **D**, **E**, **F** As for **A**, **B**, **C** but where a low-energy vortex distribution is injected by a grid of narrow circular barriers

two stirs in comparison with simulations. The vortex number for the paddle stir shows almost complete suppression of vortex decay over 10 s, indicating a strong spatial segregation of oppositely-signed vortices. In contrast, the grid stir loses 60% of the vortices in this time to vortex annihilation and edge losses. Figure 6.3F plots the vortex nearest-neighbour distance ℓ/ℓ_0 (where $\ell/\ell_0 \simeq 1$ indicates a uniform distribution). While this quantity increases with time for the clustered state, indicating spreading of the clusters, it remains <1 for the entire 10 s duration. By contrast, for the grid stir, ℓ/ℓ_0 stays quasi-constant and near unity, characteristic of a disordered state.

In the clustered phase the simulations demonstrate that vortex signs can be dependably inferred for $t \leq 5$ s from the experimental positions of the vortices relative to the minor axis of the ellipse (see Sect. 6.4.8). From these data, we can estimate the energy of the experimental vortex configurations as a function of time using the point-vortex model, including boundary effects (see Sect. 2.3.2), and compare with GPE simulations, as shown in Fig. 6.3G. Despite a gradual decay of the energy,

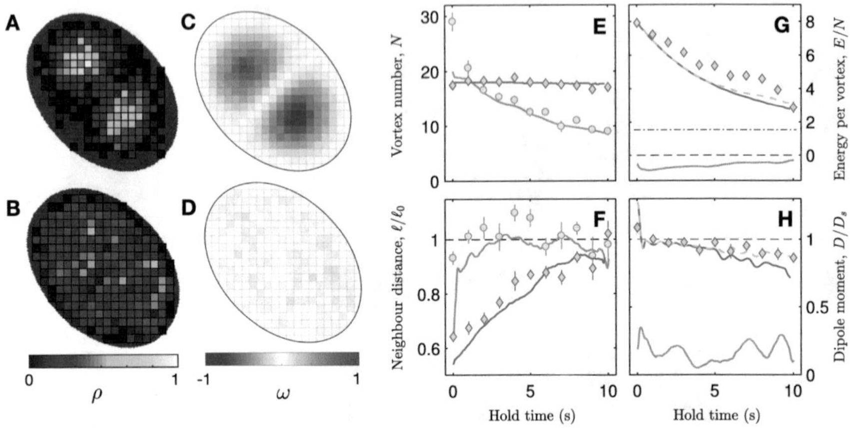

Fig. 6.3 Evidence of vortex cluster metastability. Experimental (unsigned) vortex density histograms $\rho = \sigma^+ + \sigma^-$ for **A** paddle and **B** grid stirs, respectively. The data are collected following hold times of $t = \{0, 1, 2 \ldots 10\}$ s, with 10 samples at each time (110 samples total). **C, D** Corresponding GPE simulation (signed) vorticity histograms $\omega = \sigma^+ - \sigma^-$ time averaged over 0–10 s. **E, F** Experimental average vortex number $\langle N \rangle$ and nearest neighbour distance ℓ/ℓ_0 versus hold time, where $\ell_0 = \sqrt{0.89\pi ab/N}$ is the expected value for a uniform distribution within the 89% detection region of the a:b ratio ellipse: paddle sweep (blue diamonds) and grid sweep (orange circles). GPE simulation results are shown as solid lines of the same colour. **G** Point-vortex energy versus time. Blue diamonds: experimental estimate. Blue solid line: exact point vortex energy from GPE. Blue dashed line: estimate from applying the experimental protocol to the GPE data. The black horizontal dotted line indicates the energy of the state with $T = \pm\infty$ and the purple dash-dotted line indicates $T = T_c$. The orange line indicates the energy of the grid-sweep simulation. **H** Dipole moment versus time. Lines and markers are as in **G**. Red dashed line shows the supercondensate limit $D = D_s$

the system remains well within the negative temperature region for the entire 10 s hold time, equivalent to approximately 50 times the initial cluster turnover time of $\sim$0.2 s. The decay is due to a combination of the finite lifetime of the condensate ($\tau = 28 \pm 2$ s), residual thermal fraction of $\sim$30%, and residual non-uniform BEC density of $\sim$6% RMS. The non-uniformity of the density is due to harmonic nature of the sheet. This conclusion is supported by GPE simulations with phenomenological damping, which are in agreement with experimental observations (see Fig. 6.3G). The grid sweep simulation shows a small increase in energy per vortex over the hold time, indicating that evaporative heating marginally prevails over thermal dissipation; annihilations manage to drive the system towards the negative temperature region, but not into the clustered phase.

Similarly, we may estimate the dipole moment, $D \equiv N^{-1} \left|\sum_i \mathrm{sgn}(\Gamma_i)\, \vec{r}_i\right|$ and the vortex temperature T, which we compare with analytic mean-field theory predictions [14]. The average dipole moment is the order parameter for the clustering transition; above the clustering energy, it begins to increase monotonically with energy, and it approaches an asymptotic maximum D_s in the supercondensation limit. For our elliptical geometry $D_s \simeq 0.47a$, for semi-major axis a (see Sect. 6.4.7).

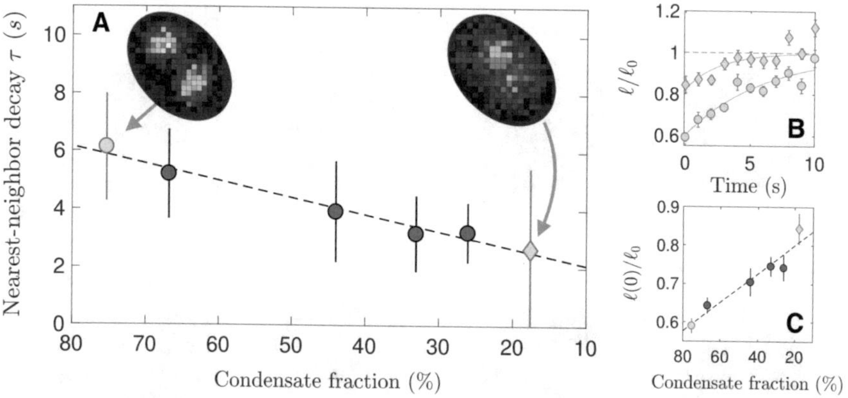

Fig. 6.4 Cluster decay rate versus BEC fraction. **A** Decreasing condensate fraction results in more rapid cluster dissociation, indicated by decreasing nearest-neighbour distance decay times as determined by exponential fits, **B** here showing the nearest-neighbour distance decay for the largest (blue circles) and smallest (green diamonds) condensate fractions. Insets show time-averaged vortex density histograms accumulated over a 10 s hold for these cases, as in Fig. 6.3; see Fig. 6.6 for the full set of time-averaged histograms, and Fig. 6.5 for the histograms immediately following the sweep. **C** The initial nearest-neighbour distance $\ell(0)$ increases with decreasing condensate fraction, indicating limitations in injecting high-vortex energy in the presence of thermal damping. Dashed lines indicate linear fits to the data

Figure 6.3H shows that the paddle stir exhibits a large dipole moment, with an average of $D/D_s \sim 89\%$ over the 10 s hold time. The experimental estimate agrees well with simulations for $t \leq 5$ s, when oppositely signed vortices remain completely segregated on opposite sides of the minor axis. By contrast, for the grid stir $D/D_s \sim 1/\sqrt{N}$ (c.f. Fig. 6.3E), consistent with an unclustered phase at finite N [14]. Finally, mean-field analysis shows that the clustering transition occurs at a temperature $T_c \simeq -0.31 T_0 N$ (see Sect. 6.4.8), while supercondensation [14] occurs at $T_s = -0.25 T_0 N$ (see Fig. 6.1F). We estimate the final temperature from the point-vortex energy, finding $T_{exp} \simeq -0.28 T_0 N$.

Thermal friction is expected to play a major role in the damping of the Onsager vortex clusters [31]. We experimentally investigated the role of an increased thermal component by injecting clusters for a range of smaller condensate fractions (i.e., higher BEC temperatures), while maintaining similar injected vortex number (Fig. 6.7). As shown in Fig. 6.4A, with decreasing condensate fraction we observe a reduction of the nearest-neighbour distance decay time to the uniform value $\ell/\ell_0 \simeq 1$, obtained by empirical fits (see Sect. 6.4.10), (examples in Fig. 6.4B); cumulative vortex histograms for the largest and smallest condensate fractions (insets) also show diminished clustering with decreasing condensate fraction. Furthermore, the initial nearest-neighbour distance increases with decreasing condensate fraction (Fig. 6.4C), suggesting the injection of high energy is less efficient in the presence of strong damping. These results suggest thermal dissipation is more important than losses to sound in our experiment; indeed, Gross-Pitaevskii simulations without ther-

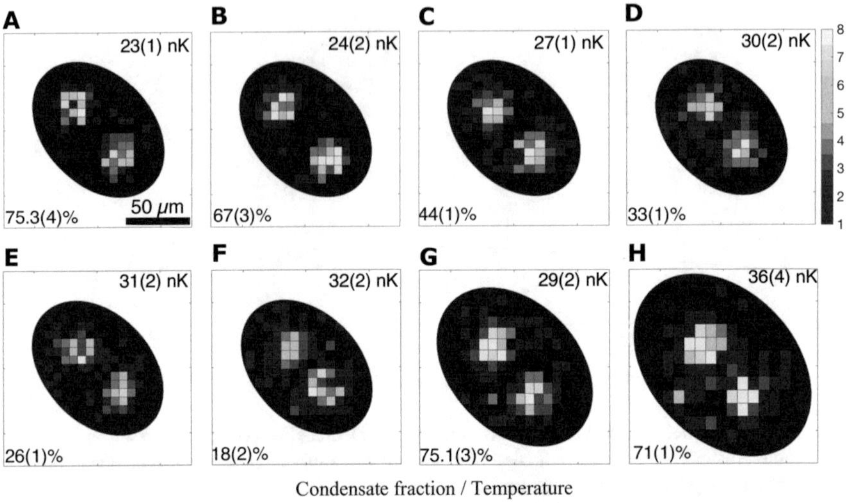

Fig. 6.5 Initial position vortex histograms immediately after the sweep. **A–F** Vortex position histograms corresponding to the full condensate fraction and temperature range in the $\{2a, 2b\} = \{120, 85\}$ μm trap considered in Fig. 6.13. **G, H** Vortex position histograms for the $\{140, 100\}$ μm trap, and the $\{160, 115\}$ μm trap. The initial condensate fraction is indicated in the bottom left, and the temperature in the top right of each subfigure. The temperatures and condensate fractions were measured using TOF

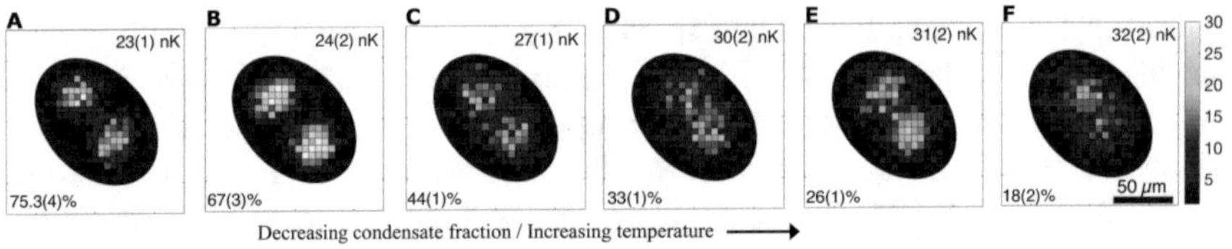

Fig. 6.6 Time-averaged vortex position histograms as a function of condensate fraction. **A–F**, Vortex position histograms corresponding to the full condensate fraction and temperature range in the $\{2a, 2b\} = \{120, 85\}$ μm trap considered in Fig. 6.13. The initial condensate fraction is indicated in the bottom left, and the temperature in the top right of each subfigure

mal damping (thus containing only losses from vortex-sound coupling) were found to support this conclusion as the clusters retained over 90% of their initial energy. Thermal friction may limit future experiments from observing the dynamic emergence of Onsager vortex clusters.

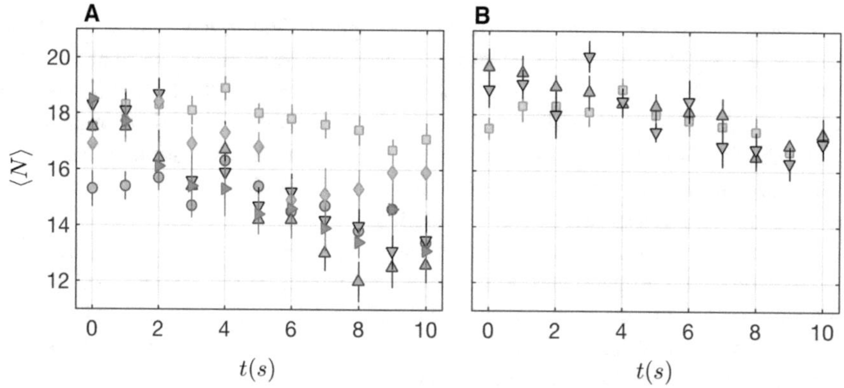

Fig. 6.7 Mean vortex number versus hold time for the paddle stir. **A** Vortex numbers as a function of time for different condensate fractions: 75.3(4)% (red circles), 67(3)% (green squares), 44(1)% (blue upward facing triangles), 33(1)% (black downward facing triangles), 26(1)% (orange diamonds), 18(2)% (red right facing triangles). **B** Vortex numbers as a function of time for the larger traps, with larger condensate density variations: $\{2a, 2b\} = \{120, 85\}$ µm trap and 67(3)% fraction (green squares), $\{140, 100\}$ µm trap (blue upward triangles), and $\{160, 115\}$ µm trap (black downward triangles)

6.4 Experimental Methods

6.4.1 Optically Configured Bose-Einstein Condensates

Our experimental apparatus is described in Chap. 3. Here we summarize the important parameters of the system for the presented experiment. The ^{87}Rb BEC is confined in a red-detuned laser sheet, providing harmonic trapping in the vertical z dimension with frequency $\omega_z = 2\pi \times 108$ Hz. The trapping in the x-y plane can be arbitrarily configured via direct projection of blue-detuned light which is patterned with a digital micromirror device (DMD) as described in Chap. 4. The BEC is formed using a hybrid optical and magnetic trapping technique [32]. We initially evaporate in a hybrid trap produced from a single, radially symmetric 95 µm waist 1064 nm red-detuned Gaussian beam and relaxed quadrupole magnetic field. Before reaching the BEC critical temperature, we transfer the atoms to a 1064 nm red-detuned Gaussian sheet and simultaneously ramp the magnetic field to approximately zero. Optical evaporation over four seconds produces BECs of up to $N_{\text{tot}} = 3 \times 10^6$ atoms in the approximately azimuthally symmetric harmonic optical trap with $\{\omega_x, \omega_y, \omega_z\} = 2\pi \times \{6.8, 6.4, 360\}$ Hz. In the final second of evaporation, the 532 nm light illuminating the DMD is linearly ramped to a peak value of 10μ, where $\mu = k_B \cdot 22$ nK is the chemical potential, producing a highly-oblate configured BEC with $N_{\text{tot}} \sim 2.2 \times 10^6$ and 67(3)% condensate fraction in a 125 µm × 85 µm hard-walled elliptical trap. With the optical trapping beams held on, we levitate the cloud against gravity by ramping on an unbalanced quadrupole magnetic field, which additionally results in an 80 G DC residual magnetic field in the vertical direction. Simultaneously, we reduce the sheet trapping power resulting in the final trap frequencies $\{\omega_x, \omega_y, \omega_z\} \sim 2\pi \times \{1.8, 1.6, 108\}$ Hz, and trap depth of ~90 nK. We also reduce the DMD pattern depth to $\sim 5\mu$. Combined with the hard-walled confinement of the DMD, this results in an approximately uniform atom distribution with a calculated vertical Thomas-Fermi diameter of 6 μm and a healing length of $\xi \sim 500$ nm at the centre of the trap (average $\xi \sim 530$ nm).

We measure the BEC lifetime in this trap to be ~28 s, which is shorter than the vacuum-limited lifetime of ~60 s. This suggests that scattering from the optical trap is a source of atom loss. We expect that a trap based on blue-detuned light would reduce this loss, and lead to increased condensate fraction which could potentially increase the lifetime of the vortex clusters.

6.4.2 Obstacle Sweeps

The paddle and grid obstacles are formed using the DMD. To dynamically alter the potential we upload multiple frames to the DMD, with the initial frame being the empty elliptical trap. Elliptically-shaped paddles, with a major and minor axis of 85 μm and 2 μm respectively, are then swept through the BEC at constant velocity. The paddle sweeps are defined by a set of 250 frames and the barriers start slightly outside the trap edge, with the paddles intersecting the edges of the elliptical trap at their midpoints. A 150 μm s^{-1} sweep (~0.1 c, where the speed of sound $c \sim$ 1290 μm s^{-1}) is utilised for the $\{2a, 2b\} = \{120, 85\}$ μm trap, which results in a sweep time of 580 ms. Sequential paddle positions are separated by ~350 nm, resulting in sufficiently smooth translation. After crossing the halfway point, the paddles are linearly ramped to zero intensity by reducing the major and minor axes widths to zero DMD pixels. For the grid case, an array of seven 4.5 μm diameter barriers was swept at the increased velocity of 390 μm s^{-1}, due to a higher critical velocity for vortex shedding. The barriers heights were then linearly ramped to zero after crossing the halfway point.

6.4.3 BEC Imaging and Vortex Detection

For darkground Faraday imaging (see Sect. 2.5.3), we utilize light detuned by 220 MHz from the ^{87}Rb $|F = 1\rangle \rightarrow |F' = 2\rangle$ transition in an 80 G magnetic field with 52.6× magnification. This results in images with a measured resolution of 1080(15) nm FWHM at 780 nm illumination (see Sect. 4.4.2). For the small phase shifts imparted by our vertically-thin cloud, raw Faraday images return a signal $\propto n_0^2$ which improves vortex visibility through exaggerating density fluctuations. The density can be determined through postprocessing (see Sect. 2.5.3). The $\xi \sim 500$ nm healing length results in poor vortex visibility in situ, see Fig. 6.2A, D. However,

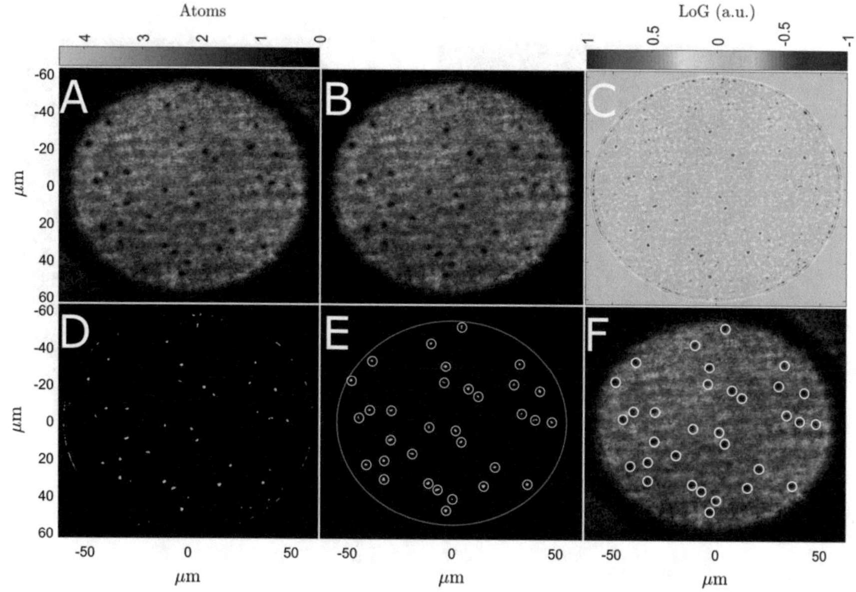

Fig. 6.8 Vortex position detection. **A** Absorption image of the experimental BEC column density. **B** Selected region of interest for vortex position detection. **C** The Gaussian blob algorithm takes the Laplacian of the Gaussian-filtered image. **D** A threshold is applied to the signal to create simply connected binary clusters. **E** The clusters are sorted by size and Euler's number and the center of the vortex with the right parameters are assumer to be the vortex location. **F** Absorption image of the experimental BEC column density with indicated vortex positions

Faraday imaging combined with a short 3 ms time of flight (TOF), where the optical beams are suddenly turned off, and the levitation field is held on, improves the vortex visibility significantly, while the column density is otherwise essentially unchanged, see Fig. 6.2B, E. After masking the image with the elliptical pattern, vortices are detected automatically using a Gaussian blob vortex image processing algorithm [33] that examines connected regions of a thresholded background-subtracted image, see Figs. 6.8 and 6.9. We restrict detection of vortices to the inner 89% of the ellipse area to avoid spurious detections near the condensate edge.

6.4.4 *Effective 2D Theory*

We use the dimensional reduction technique presented in Sect. 2.2.3, with a harmonic Thomas-Fermi profile to determine a 2D GPE for the system. For a BEC of $N_{\mathrm{tot}} = 2.25 \times 10^6$ atoms in the 120 μm × 85 μm trap, we obtain $\mu_{2D}/k_B = n_0 g_2/k_B = 19.63$ nK, $\xi = \hbar/\sqrt{m\mu_{2D}} \approx 0.533$ μm and $c = \sqrt{\mu_{2D}/m} \approx 1370$ μm/s. Values for the systems with lower condensate fraction or larger trap size are obtained by a simple

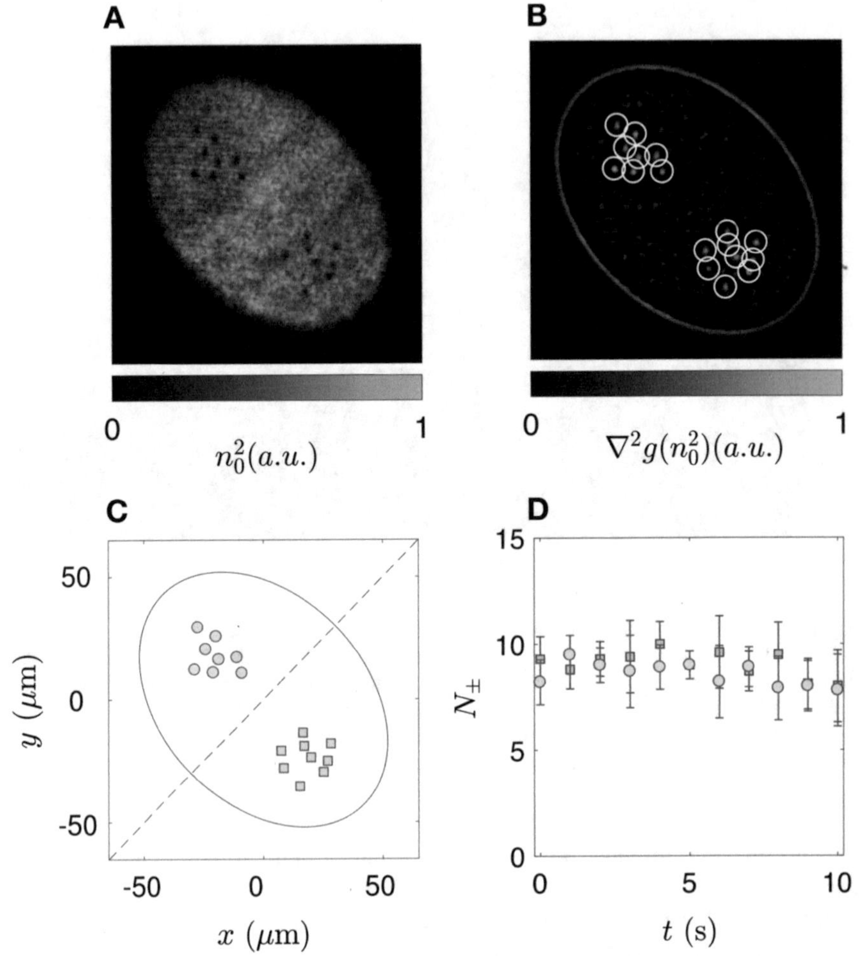

Fig. 6.9 Vortex fitting and classification. **A** Background-subtracted Faraday image of the experimental BEC column density. **B** Applying the Gaussian blob algorithm to locate the vortex cores. **C** Vortex circulations are then assigned across the minor axis, with positive vortices indicated by circles and negative vortices indicated by squares. **D** The distribution of positive and negative assigned vortices as a function of time for the data sets corresponding to Fig. 6.3A. Nearly equal numbers of positive and negative vortices are obtained throughout the hold times. The error bars indicate the standard deviation of the data

scaling. Scaling the condensate number $N_{\rm tot} \to \alpha N_{\rm tot}$ gives $\xi \to \alpha^{-1/3}\xi$ and $c \to \alpha^{1/3}c$ while scaling the trap $\{a, b\} \to \lambda\{a, b\}$ yields $\xi \to \lambda^{2/3}\xi$ and $c \to \lambda^{-2/3}c$.

6.4.5 Dynamical Modelling

Dr. Matthew Reeves modelled the dynamical evolution of the experiment using a phenomenologically damped GPE to account for energy and atom losses [34]. The 2D GPE becomes

$$i\hbar\partial_t\phi = (1 - i\gamma)\left[-\frac{\hbar^2\vec{\nabla}^2}{2m} + V(x, y, t) + g_2|\phi|^2 - \mu(t)\right]\phi, \qquad (6.1)$$

where γ is the dissipation coefficient. Up to a noise term, Eq. (6.1) is equivalent to the simple growth stochastic Gross Pitaevskii equation (SGPE), a microscopically derived model of atomic BECs that incorporates dissipation due to interactions with a thermal component [35]. The exponential decay of the atom number, $N_c(t) = N_c(0)\exp(-t/\tau)$, for the decay constant $\tau \approx 28 \pm 2$ s, is incorporated via a time-dependent chemical potential $\mu(t)$. From Eq. (2.18) this gives $\mu(t) = \mu_{2D}\exp(-2t/3\tau)$. Empirically we find that the experimental data for the paddle stir are well matched by numerical simulations with a dissipation coefficient of $\gamma = 6.0 \times 10^{-4}$. We find a slightly larger phenomenological dissipation coefficient is required for the grid stir, $\gamma = 8.5 \times 10^{-4}$. We attribute this to the increased sound production for this case (see below). The total external potential is modelled as a combination of a stationary trap and time-dependent stirring obstacles: $V(x, y, t) = V_{\rm trap}(x, y) + V_{\rm ob}(x, y, t)$. The stationary component of the trap includes the optical dipole trap in the x-y plane, and binary DMD pattern convolved with the previously measured point spread function of the optical system [36]. The stirring obstacles are modelled by steep-walled hyperbolic tangent functions, which increase to the maximum on the scale of the healing length. The numerical simulations were performed using XMDS2 [37].

Note that while the damped GPE simulations are a reasonable approximation at high condensate fractions, as shown in Fig. 6.3, full stochastic simulations would likely be required for the high-temperature and low condensate-fraction conditions of Figs. 6.4 and 6.6. As the phenomenological damping parameters are determined a posteriori, we did not simulate these cases.

Reeves also performed undamped GPE simulations to obtain an upper estimate of the sound produced from the stirring procedures. Using the standard Helmholtz decomposition of the kinetic energy [38], we find the amount of sound produced is quite small: for the paddle stir it is $\sim$1.5% of the total kinetic energy at the end of the stir and <5% at the end of the hold time (the increase is from vortices radiating

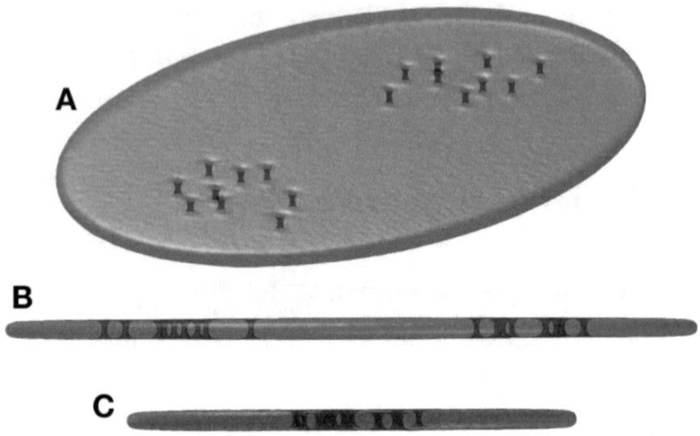

Fig. 6.10 3D GPE simulation data. Density isosurfaces (approximately 25% of peak density) are shown shortly after the paddle stir: **A** angle view, **B** view looking down the minor axis, **C** view looking down the major axis. The vortices are clearly rectilinear; no vortex bending is visible

sound as they accelerate). For the grid stir, it is ∼8% at the end of the stir and <13% at the end of the hold time.

While vortex bending becomes highly suppressed in 3D oblate harmonic traps [26], we also performed 3D GPE simulations of Eq. (6.1), shown in Fig. 6.10. Little to no vortex bending is observed, further justifying a 2D treatment of the system.

6.4.6 Calculation of Entropy

To generate the entropy versus energy curve, $S(E)$, in the elliptical domain in Fig. 6.1, Dr. Matthew Reeves generated 10^9 uniformly random, neutral configurations of $N_\pm = 9$ vortices within the ellipse, and calculate the energy for each state via Eqs. (2.56) and (2.57). The distribution are then binned to approximate the density of states, $W(E) = \int (\prod_{i=1}^{N} d^2\vec{r}_i)\delta(E - H(\{\vec{r}_i\}))$, which determines the entropy $S = k_B \log W$, where k_B is Boltzmann's constant. The energy is calculated from the vortex locations in terms of the 2D healing length ξ. Note that the energy in Eq. (2.56) is defined up to an arbitrary additive constant. To simplify the comparisons between different data sets, $E = 0$ is set to correspond to the point of maximum entropy ($T = \infty$), such that negative (positive) energies correspond to positive (negative) temperatures in all cases.

6.4.7 Upper Bound of the Dipole Moment

Considering two point vortices with opposite circulations in the $\{2a, 2b\} = \{120, 85\}$ μm elliptical domain, the mechanical equilibrium condition reads

$$0 = \kappa_i \frac{dx_i}{dt} = \frac{\partial H_\Omega}{\partial y_i}; \quad 0 = \kappa_i \frac{dy_i}{dt} = -\frac{\partial H_\Omega}{\partial x_i}, \tag{6.2}$$

where $i = 1, 2$, giving that $y_1 = y_2 = 0$ and $x_1 = -x_2 = d$ is the unique stationary point where the forces on each vortex due to the other vortex and the image vortices all cancel. Solving numerically Eq. (6.2), we find that $d \simeq 0.47a$. The value of d, depending on the geometry of the domain, sets the upper bound of the blue average dipole moment.

6.4.8 Onset of the Clustering

For an incompressible flow, one can introduce a stream function ψ to describe the flow, connected to the vorticity ω via $-\vec{\nabla}^2\psi = \omega$. The self-consistent equation of the stream function of a system containing a large number of vortices in a bounded domain Ω is [2]

$$-\vec{\nabla}^2\psi = \frac{n_0}{2}\exp(-\tilde{\beta}\psi) - \frac{n_0}{2}\exp(\tilde{\beta}\psi), \tag{6.3}$$

where $n_0 = 2/A$ is the normalized vortex number density inside the area A , and $\tilde{\beta} \equiv (E_0 N/2k_B T)$ is the inverse temperature in the natural energy units of the vortex system. Linearising Eq. (6.3) around the uniform state of vortices $\psi = 0$, the fluctuation $\delta\psi$ satisfies

$$(\vec{\nabla}^2 + \lambda)\delta\psi = 0, \tag{6.4}$$

with the Dirichlet boundary condition $\delta\psi(\vec{r} \in \partial\Omega) = 0$, here $\lambda = -4\pi\tilde{\beta}n_0$. The onset of the vortex clustering (purple star in Fig. 6.1D) occurs if Eq. (6.4) has nonzero solutions to the eigenvalue problem of the Laplacian operator in the elliptical domain [14]. In terms of elliptical coordinates Eq. (6.4) becomes Mathieu's equation. The most relevant eigenvalue associated with the transition is $\lambda = 4h^2/(a^2 - b^2)$, where h is the first positive root of the modified Mathieu function $Mc1(m, R, h)$ with $m = 1$ and $R = \tanh^{-1}(b/a)$. The transition happens at $\tilde{\beta} = \tilde{\beta}_c = -\lambda/(4\pi n_0) \simeq -1.614$, giving $T_c = (k_B\tilde{\beta}_c)^{-1}E_0N/2 \simeq -0.31T_0N$, with $T_0 = E_0/k_B$.

6.4.9 Non-uniformity of Clustered Vortex States

As noted in the Sect. 6.3, a key feature of the clustered states is that they are *non-uniform*, not only in the vorticity field $\omega(\vec{r}) = \sigma^+(\vec{r}) - \sigma^-(\vec{r})$, but also in the vortex density, $\rho(\vec{r}) = \sigma^+(\vec{r}) + \sigma^-(\vec{r})$. This contrasts with unclustered states, which exhibit *uniform* vortex density.

Figure 6.3A shows that the vortex density histogram remains non-uniform for the entire hold time for the paddle stir. This can only happen if the vortex energy is sufficiently high, since at low energies vortices are free to roam throughout the entire system. The complete absence of vortex number decay for the paddle stir (Fig. 6.3E) further supports this conclusion, as this requires the spatial segregation of

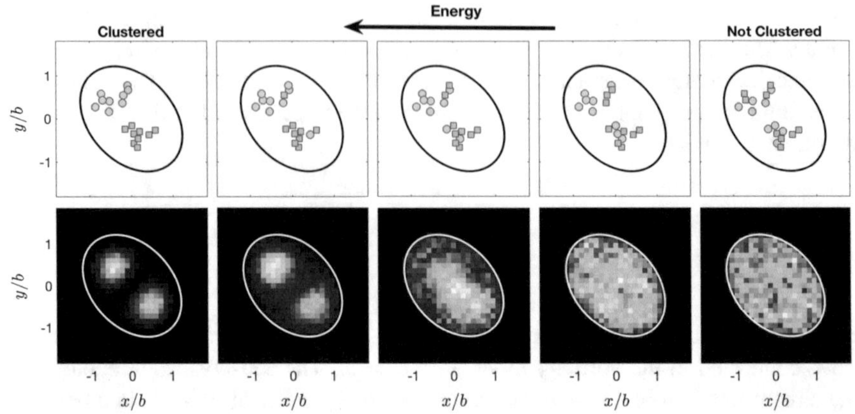

Fig. 6.11 Point-vortex dynamics. Time averaged vortex densities produced from point vortex dynamics (bottom row), using a sample from the paddle experiment for the initial positions (top row). The leftmost example assumes the clusters are all of the same sign, whereas the rightmost assumes the charges are random. Intermediate cases show the effect of selecting 1, 2, or 3 vortices from each cluster at random and swapping their signs to reduce the cluster net charge and lower the energy. Note that the initial positions are identical for all initial conditions; only the vortex signs are different

positive and negative vortices which can only occur in a high energy configuration. By contrast, the grid stir exhibits significant vortex decay (because vortex-antivortex pairs are present), Fig. 6.3E, and the density is uniform, Fig. 6.3C.

To strengthen the argument above, Dr. Matthew Reeves also consider average vortex densities at different energies under point vortex evolution via Eqs. (2.56)–(2.59). In Fig. 6.11 time-averaged vortex densities produced from 2D point vortex dynamics is shown, using a sample from the paddle experiment for the initial vortex positions. The energy can be altered by changing some of the vortex signs (while maintaining $N_+ = N_- = N/2$) for the same vortex position data. Assuming the opposite charges are completely segregated (far left) yields a histogram consistent with the experimental observations, whereas random charges (far right) instead yields constant density, as is observed for the low-energy grid stir. Only the left two panels resemble the experimental data in Fig. 6.3A. The simulations producing Fig. 6.11 did not contain any damping. However, dissipation would further smear the distributions, making a argument stronger in favour of near complete clustering.

While the Bragg-scattering procedure for sign detection [24, 30] is possible for our system, the non-uniform nature of the clustered states means we do not require sign detection for the majority of the analysis presented. Nonetheless, Bragg-scattering data was used to experimentally ensure the paddle experiment does indeed initially inject single-sign clusters. An example is shown in Fig. 6.12.

In order to determine the most likely vortex configuration, we calculate the standard deviation (standard error) between our experimentally observed Bragg-scattering differential signal, Fig. 6.12C, and a simulated Bragg differential signal.

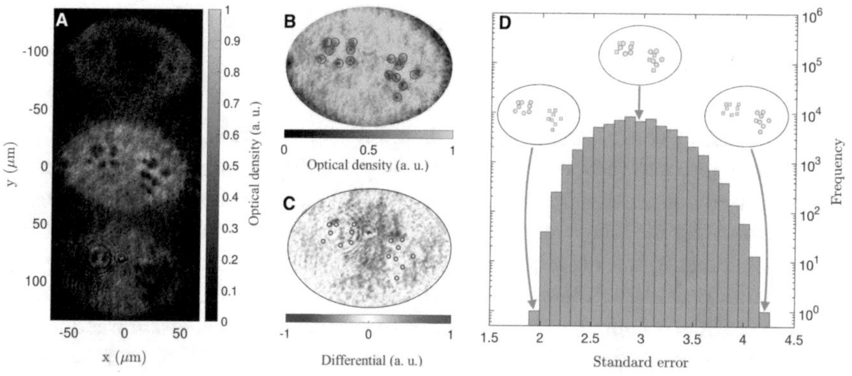

Fig. 6.12 Bragg-scattering vortex sign detection. **A** Absorption imaging of the condensate after cluster injection and Bragg scattering, followed by a 10 ms TOF, showing the Bragg-scattered components and central unscattered cloud. **B** Vortex detection on the sum of unscattered and scattered components; the corresponding differential signal of the scattered components is shown in **C**. **D** Vortex signs are permuted through the 2^{16} charge configurations for the 16 vortices (examples shown in insets), and the standard error between the experimental, **C**, and a synthesized differential signal is calculated (see text); the histogram demonstrates that a single configuration minimizes the error, containing single sign clusters as in Fig. 6.2C. Swapping the vortex signs results in the standard error being maximized. See Sect. 6.5 for detailed explanation of sign detection

The simulated signal starts with a point-vortex velocity field based on the experimental vortex positions, Fig. 6.12B, where the circulations of the vortices can be iterated continuously between $\pm\Gamma$. By using this velocity field, along with the experimentally measured Bragg-scattering response function and total BEC density, a simulated differential signal can be generated for any configuration of vortex circulations. A steepest-descent method is used to minimize the standard error between the measured differential signal and simulated profile, with the initial guess being zero circulation for all vortices. The algorithm determines that a fully polarized vortex distribution with two same-sign clusters minimizes the standard error. To confirm that the steepest-descent converges to the global minimum, Fig. 6.12D displays a histogram of all the possible 2^{16} vortex configurations versus standard error. While this second approach confirms the configuration for which the standard error is minimized, it has the disadvantage of needing to iterate through all 2^N possible vortex-sign permutations. The sign detection procedure is described in Sect. 6.5, with the example image analysed being the same as Fig. 6.12A.

6.4.10 Nearest-Neighbour Distance and Energy Decay for Varying BEC Fraction and Density

For investigating vortex cluster energy damping as a function of BEC fraction, we reduce the depth of the evaporative cooling ramp, leading to an increased temperature and decreased condensate fraction. During the levitation procedure, we find loss of thermal atoms for the hotter conditions which occurs at a rate inefficient for continued evaporation, due to the reduction in the optical dipole intensity and corresponding reduction in trap depth. This results in an approximately constant $N_{\text{tot}} \sim 3.3 \times 10^6$ atoms in the final potential, while the final temperature and condensate fraction vary. The full range of temperatures and condensate fractions utilized were $T = \{23(1), 24(2), 27(1), 30(2), 31(2), 32.7(2)\}$ nK and $N_c/N_{\text{tot}} = \{75.3(4), 67(3), 44(1), 33(1), 26(1), 18(2)\}\%$, respectively.

Gross-Pitaevskii equation (GPE) simulations have previously shown that increasing non-uniformity in the density of the condensate inhibits the dynamic formation of Onsager vortices [25]. To determine the sensitivity of our experiment to non-uniform density, we have increased the variation in the density of our BEC before performing the paddle sweep. This is achieved by increasing the size of our trap while maintaining its aspect ratio, which increases the relative contribution of the residual harmonic optical trap from the optical dipole sheet potential to the 2D confinement. We thus examined three different trap sizes, $\{2a, 2b\} = \{125, 85\}$ µm; $\{140, 100\}$ µm; $\{160, 115\}$ µm with RMS density variation of $\Delta n_0 = \{6.2\%, 8.1\%, 10.9\%\}$, respectively. For the larger elliptical traps, the paddle sizes are proportionately scaled. A 150 µm s^{-1} paddle sweep is maintained for the $\{140, 100\}$ µm trap, but a 136 µm s^{-1} velocity is used to produce a similar number of vortices for the $\{160, 115\}$ µm trap. For the $\{140, 100\}$ µm trap, the temperature and condensate fraction was $T = 29(2)$ nK and $N_c/N_{\text{tot}} = 75.1(3)\%$, while the $\{160, 115\}$ µm trap had $T = 36(4)$ nK and $N_c/N_{\text{tot}} = 71(1)\%$.

As the total atom number N_{tot} is approximately constant for all conditions, varying the BEC fraction and trap size leads to varying healing lengths, which we scale appropriately for calculating the vortex energy and nearest-neighbour distance. We furthermore determine the density of states $W(E)$ for $N_\pm = 9$ vortices to determine the peak value for each healing length and shift the energies as described in previous sections.

For the nearest neighbour decay shown in Figs. 6.4, 6.13, we fit an empirical exponential decay function $\ell(t)/\ell_0 = ae^{-t/\tau} + 1$, where the limiting value $\ell/\ell_0 \simeq 1$ is expected for uniformly distributed vortices. The resulting variation in nearest-neighbour distance decay times when varying the density is shown in Fig. 6.13A, which, in contrast to Fig. 6.4, shows little variation in the decay rate, and reduced variation in the initial nearest-neighbour distance, Fig. 6.13C.

We also apply the energy estimation procedure to the data; we fit the energy decay with the empirical function $E(t)/N = ae^{-t/\tau} + h_0$, where h_0 is a constant determined by a preliminary fit. To test the reliability of inferring the vortex energy based solely on vortex positions, we have also numerically generated random (unclustered) ensembles of $N_\pm = 8$ vortices, equal to the mean number of vortices detected

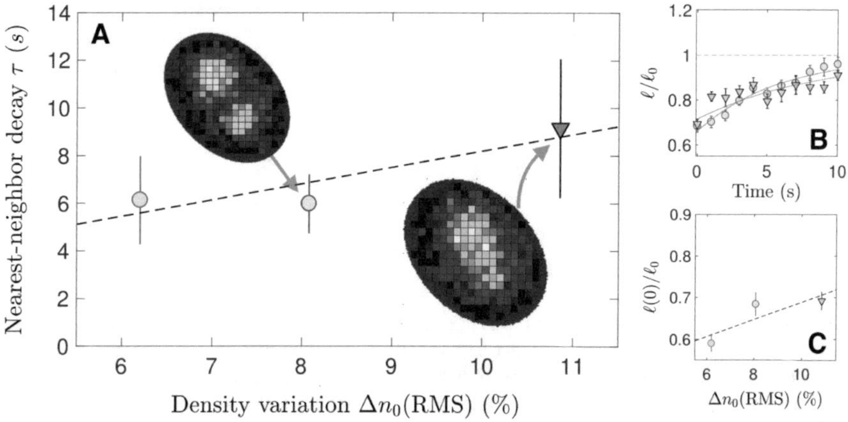

Fig. 6.13 Cluster decay rates with increased density variation. **A** Increasing the ellipse size results in the residual harmonic confinement becoming more significant, leading to increased density variation, while maintaining high condensate fractions. Nearest-neighbour distance decay times are determined by exponential fits (see text), **B** with the $\{2a, 2b\} = \{125, 85\}$ μm trap (blue circles), {140, 100} μm trap (red circles), and largest {160, 115} μm trap (grey triangles) shown. Insets show time-averaged vortex density histograms accumulated over a 10 s hold for larger cases, as in Fig. 6.4; see Fig. 6.6 for the full set of time-averaged histograms, and Fig. 6.5 for the histograms immediately following the sweep. **C** The initial nearest-neighbour distance $\ell(0)$ varies over a smaller range when compared with Fig. 6.4. Dashed lines indicate linear fits to the data

within the 89% detection region. We then calculate the energy of the configuration assuming that all vortices on the top left (bottom right) half of the ellipse have positive (negative) circulation. This energy is indicated in the insets of Fig. 6.14A, B as horizontal shaded regions, representing a 95% confidence interval corresponding to the 10 samples at each hold time ($\pm 1.96\sigma/\sqrt{10}$). For the times the experimentally estimated vortex energy is larger than this value, we are confident that the vortices remain clustered despite the loss of vortex energy.

We observe a sharp reduction in the cluster energy decay time for condensate fractions below 65% (Fig. 6.14A), which along with the nearest-neighbour distance analysis, Fig. 6.4, confirms thermal friction as a primary source of dissipation. We also find that increasing the density variation leads to apparent increased energy damping, as shown in Fig. 6.14B, in contrast to the nearest-neighbour distance behaviour in Fig. 6.13. While suggesting some increased energy loss due to the increased density variation, we note that the nearest-neighbour distance behaviour indicates relatively tight vortex clusters are maintained. We note that the decay fit for the largest {160, 115} μm trap (Fig. 6.14B inset) tends towards an energy value above the uncorrelated estimate. In conjunction with the vortex position histogram shown in Fig. 6.14A, we speculate that this may indicate the emergence of a monopole state, consisting of a central like-circulation cluster surrounded by opposite circulation vortices and possessing net angular momentum. In a weakly elliptical trap, this state will have a comparable entropy to the (maximal entropy) dipole configuration [19, 39].

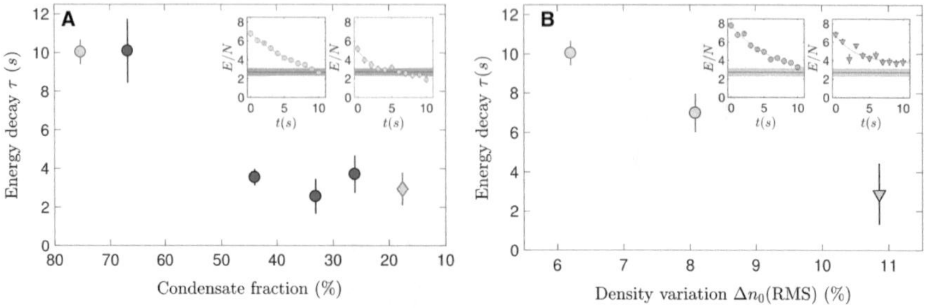

Fig. 6.14 Energy damping rates for varying BEC fraction and non-uniform density. **A** Energy decay times for varying BEC fraction, displaying a decrease in damping time with increased thermal fraction, consistent with nearest-neighbour distance and histogram analysis. (Insets) Energy versus hold time for the largest (blue circles) and smallest (green diamonds) condensate fractions. **B** Energy decay times with increasing non-uniform density, where the leftmost point corresponds to the $\{2a, 2b\} = \{120, 85\}$ μm trap. (Insets) The decay of the vortex energy for the intermediate (red circles) and largest (black triangles) traps. The shaded region indicates the upper bound of the vortex configuration energy if our circulation allocation algorithm was applied to a random vortex ensemble (see text). The lifetime is determined by fits to offset exponential decays, shown with dash-dot lines

6.5 Determining Vortex Signs

One of the important factors when looking at turbulence in a BEC is determining the sign of the singly-charged vortices in the condensate. In the limit where 2D point vortex descriptions are valid, vortex positions and signs allow one to determine all the parameters of interest. There have been many proposed methods to detect the sign of vortices [40]. Although these methods are theoretically sound, they can be technically challenging to implement. Recently, a relatively simple method using double-Bragg spectroscopy for the detection of the signs of the vortices has been demonstrated by Yong-il Shin's group at Seoul National University [24]. We implemented this detection method in our experimental. This technique relies on velocity components parallel to the Bragg beam having different coupling efficiencies (see Sect. 2.5.4), along with the superfluid flow near a vortex being mostly determined by the vortex circulation. It also relies on the Bragg recoil velocity being much higher than the highest flow in the condensate, such that most displacement is due to the Bragg scattering and not the initial in-trap atom velocity. Under these conditions, the scattered atoms spatial profile will reflect the in trap incompressible flow velocity profile.

In this section, the steps taken to detect the sign of vortices are described in detail. These steps can be divided into three main categories. First the scattering of the initial atomic cloud using double Bragg beams is described. This is followed by a description of the image processing used to detect the scattered components, the Bragg beams propagation direction, and the vortex locations. From this information, the most likely vortex signs can be found by minimizing the standard error between the measured scattered signal and a synthesized signal for a given vortex distribution.

6.5.1 Imaging Vortices

To perform Bragg spectroscopy and image the vortices in our set-up, we use light tuned to the $5^2S_{1/2}|F = 2\rangle \rightarrow 5^2P_{3/2}|F' = 2, 3\rangle$ transition with a detuning of $\delta/2\pi = 280$ Hz from the two-photon Bragg resonance. Before performing the Bragg kick, we levitate the cloud for 300 μs after release from the optical trap then pulse the Bragg beam for 1.2 ms. It takes about 5 ms for the scattered components to separate from the stationary component. This separation time also allows for the vortex cores to expand making them easy to resolve. Once the clouds are separated, an absorption image of the cloud is taken example shown in Fig. 6.12A.

6.5.2 Recognising Bragg-Scattered Components

The recognition of the position of the scattered and unscattered components is done with a modified version of the image recognition algorithm introduced in Sect. 4.5.1. This is divided into two steps, first, the recognition of the unscattered density components, and second, the recognition of the scattered components through the process depicted in Fig. 6.15. After the scattered image is taken (Fig. 6.15A), we use the the DMD pattern as the best initial guess of the unscattered BEC component shown in Fig. 6.15B. Through image recognition the initial guess image overlap is optimized allowing for rotation and translation during image recognition to find the unscattered components location. Figure 6.15C, D show the un-optimized and optimized overlaps respectively.

Once the exact position of the unscattered component has been determined, it can be removed from the image, which allows for the determination of the position of the scattered cloud components, Fig. 6.15E. As the best guess for the scattered density profile we use the unscattered density image allowing for symmetric translation from the initial position of the unscattered component through the free parameters of displacement length (l) and Bragg angle (θ) as shown in Fig. 6.15F. The initial guess for the displacement length is the recoil velocity [Eq. (2.95)] multiplied by the time between the scattering event and the time at which the image was taken; whereas, the angle is measured experimentally (Note Fig. 6.15F, G uses a bad initial guess on purpose to make it easier to see the optimization and show that it converges even for poor initial parameters). Again using image recognition and mid-way renormalization at every step, we can optimize the Bragg angle and displacement as shown by going from the initial guess Fig. 6.15G to the final optimized guess Fig. 6.15H.

The double Bragg scattering fractions, as a function of detuning δ from the stationary Bragg resonance frequency, are experimentally measured as shown in Fig. 6.16.

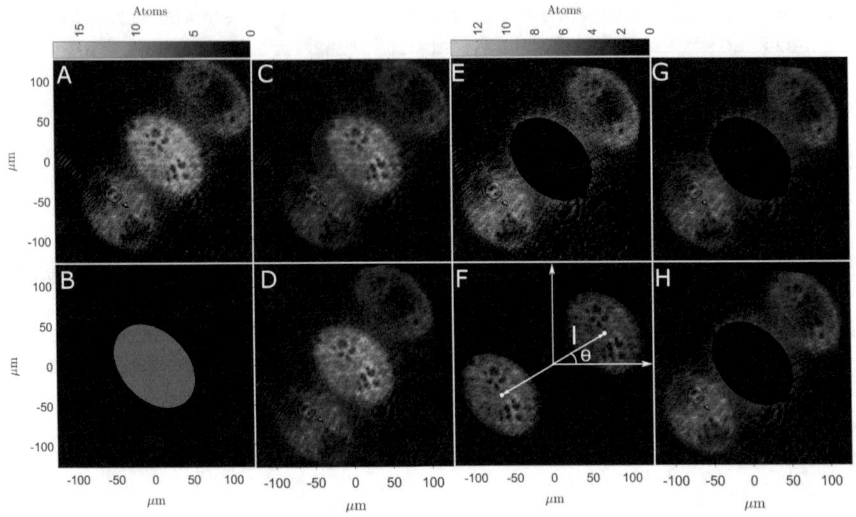

Fig. 6.15 Extraction of the scattered and unscattered components using image recognition from **A** the original atomic image. **B** DMD trapping potential which is used as the initial guess for finding the unscattered component. **C, D** Initial and final overlaps of the guess (purple) and atomic density (green) in the finding of the unscattered component which is used to establish a transformation matrix between the DMD and the camera. **E** represents the initial image with the unscattered component removed. **F** is the guess of the shape and position of the scattered components, it is estimated to be made up of displaced unscattered components with the free parameters being the scattering angle (θ) and the travelled distance (l). **G, F** represent the pre and post-optimization of the guess atomic density (purple) and the scattered density (green). Note that the initial errors were exaggerated for illustrative purposes

Fig. 6.16 Scattered atom number fractions measured for a stationary BEC as a function of δ. The squares and circles denote the atom number fractions in the two arms of the double Bragg beams. The curved line is a Gaussian best-fit to the data

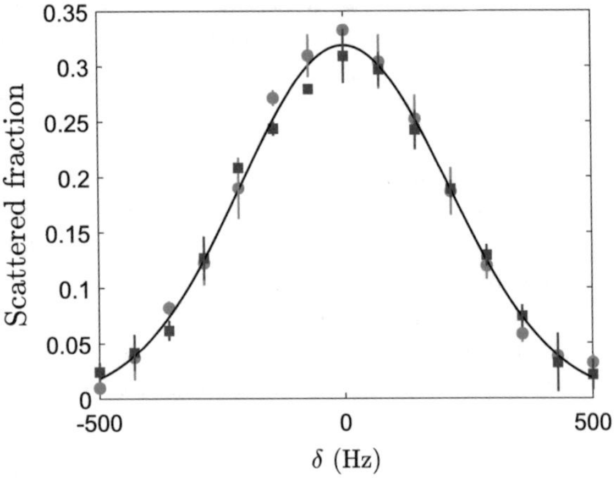

A Gaussian function is fit to the scattering response

$$\sigma(\delta) = A \exp\left[-(\delta - \delta_0)^2 / (2\delta_{\mathrm{w}}^2)\right], \tag{6.5}$$

where $A = 0.31 \pm 0.01$, $\delta_0 = 0 \pm 2$ Hz, and $\delta_{\mathrm{w}} = 209 \pm 9$ Hz. δ_{w} is the parameter used to account for the broadening due to the finite Bragg pulse length.

6.5.3 Differential and Pre-Bragg Density Extraction

Once the positions of the scattered and the unscattered components (Fig. 6.17B–D) have been determined, it is possible to reconstruct an estimate of the atomic map of the cloud assuming the scattering event had not taken place, $n_{\text{pre-Bragg}}$. This is done by summing the unscattered and scattered components, which give Fig. 6.17E and are used to convert fractional scattering maps (Fig. 6.18H) into "expected" scattered atoms, Fig. 6.18I. The differential scattered atoms, $S_{\text{measured}}(x, y)$, are obtained by subtracting the scattered components from each other (Fig. 6.17F) and are used to determine the sign of the vortices in the cloud since the different vortices.

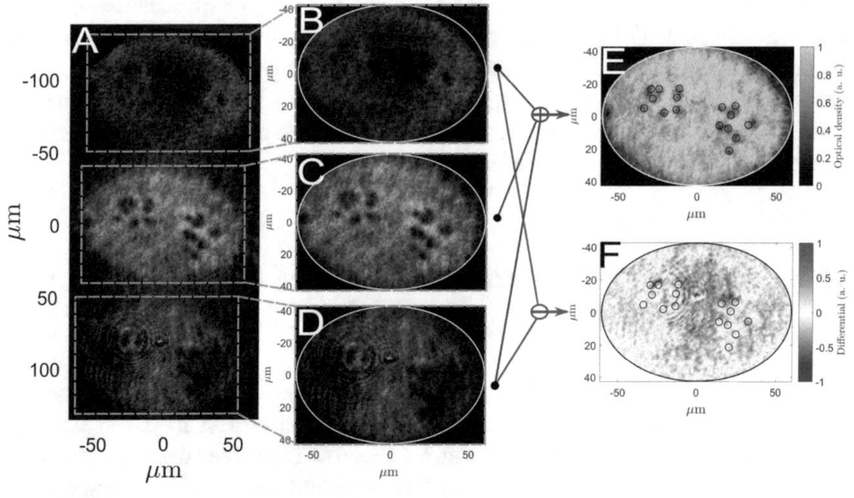

Fig. 6.17 Extraction of the pre-Bragg estimated density and differential signal. **A** Atomic density image, **B**, **D** are the extracted scattered Bragg scattered atoms, **C** is the unscattered atoms. **E** represents the best estimator of the density profile before the Bragg scattering and is obtained by summing **B** + **C** + **D**. **F** is the differential signal between the scattered atoms and is the difference between **B** and **D**. Both **E** and **F** are needed for the extraction of the vortex sign in our algorithm

6.5.4 Creation of a Differential Scattering Map

To compare with the measured differential atomic map (Fig. 6.17F), an algorithm generates differential scattered atomic maps based on the experimentally determined vortex positions from the pre-Bragg atomic map (Fig. 6.18A), by first assigning each vortex a circulation ($\vec{\Gamma}$) shown in Fig. 6.18B. Using the conformal mapping method described in (Sect. 2.3.2), the vortex positions are mapped from the oval onto the unit circle, and the locations of the image vortices are computed, (Fig. 6.18C). From these positions and assigned circulations, the velocity field ($\vec{v}_c(x, y|\vec{\Gamma})$) can be determined on the unit circle, Fig. 6.18D. Using the derivative of the conformal map determines how displacements relate to one another when transforming between the two planes as derived in Eq. (2.60). The resulting velocity field on the oval ($\vec{v}_c(x, y|\vec{\Gamma})$) is shown in Fig. 6.18E.

The component of the velocity field that couples to the Bragg beams is the one along the beam propagation direction ($\pm\hat{y}'$). The velocity field is therefore projected onto the Bragg beam propagation direction, $v_{\text{Bragg}}(x, y|\vec{\Gamma}) = \vec{v}_o(x, y|\vec{\Gamma})\hat{y}'$, resulting in Fig. 6.18F. This velocity field can be converted to a Bragg detuning ($\delta_v(x, y|\vec{\Gamma})$) using Eq. (6.5) (Fig. 6.18G). Using the measured Bragg scattering frequency response $\sigma(\delta)$, see Eq. (6.5), the Bragg detuning can be converted to a fractional scattering differential map (Fig. 6.18H) using $S_{\text{frac}}(x, y|\vec{\Gamma}) = \sigma(\delta_v(x, y|\vec{\Gamma}) - \delta_0) - \sigma(\delta_v(x, y|\vec{\Gamma}) + \delta_0)$, where δ_0 is the relative detuning of the double Bragg laser (140 Hz). Using the pre-Bragg atomic map ($n_{\text{pre-Bragg}}(x, y)$) Fig. 6.17E, the expected differential scattered atoms map can be calculated by performing the product $S_{\text{generated}}(x, y|\vec{\Gamma}) = n_{\text{pre-Bragg}}(x, y)S_{\text{frac}}(x, y|\vec{\Gamma})$. The result is shown in Fig. 6.18I which is a generated differential map that can be compared with the measured differential scattered atoms in Fig. 6.17F.

6.5.5 Determining the Maximum Likelihood Vortex Configuration

The maximum likelihood configuration of the vortices is the one whose synthesized differential scattered atoms map Fig. 6.18 is the closest to the experimental differential scattered atoms map Fig. 6.17F. How close the two maps are from one another can be calculated through the standard error (STE) between the maps which is simply the standard deviation of the difference of the differential map $STE(\vec{\Gamma}) =< (S_{\text{generated}}(x, y|\vec{\Gamma}) - S_{\text{measured}}(x, y))^2 > - < S_{\text{generated}}(x, y|\vec{\Gamma}) - S_{\text{measured}}(x, y) >^2$. One way to minimize the metric is to go through all 2^N possible vortex configurations and find the configuration which minimizes the standard error, shown in Fig. 6.19A. For this case, a fully polarized vortex distribution minimizes the metric, whereas the oppositely signed fully polarized vortex configuration maximizes the metric, and all the other possible vortex configurations fit somewhere in between. It is interesting to look at the number of mislabelled vortices as a function of the stan-

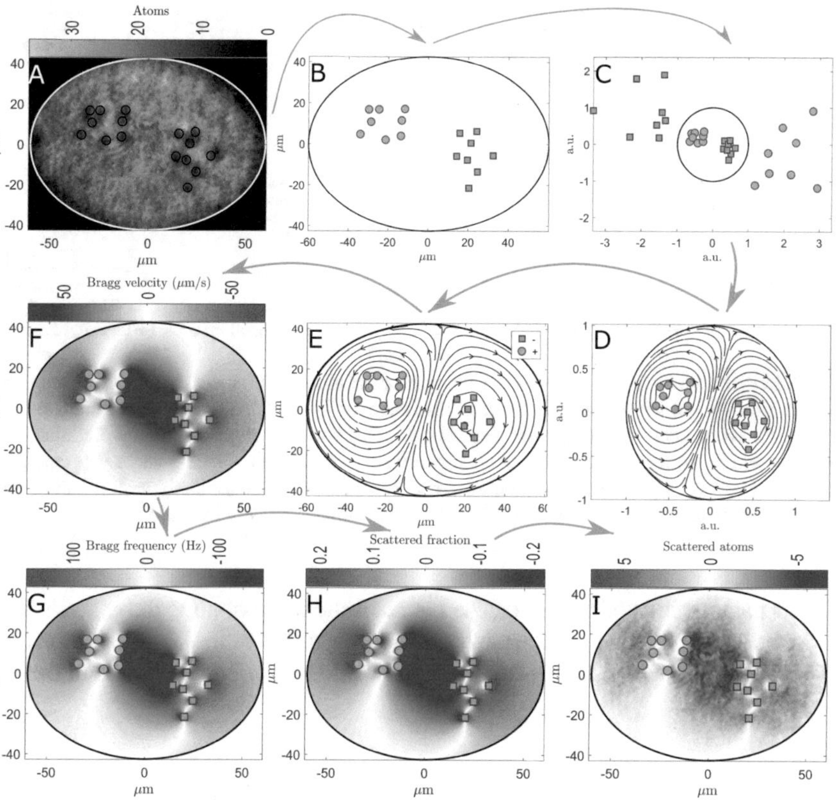

Fig. 6.18 Generation of a differential map by the sign detection algorithm. **A** the first step is to detect the initial position of the vortices in the condensate as shown in Fig. 6.9 and the extracted pre-Bragg density Fig. 6.17E. **B** Position of the vortices in the oval frame and their assigned sign. **C** Conformal mapped position of the vortices on the unit circle with their image vortices. **D**, **E** Generated flow field for assumed vortex distribution on the unit circle and mapped back onto the oval. **F** Velocity component along the Bragg pulse propagation direction. **G** Bragg frequency resonance detuning due to the velocity field. **H** Expected differential scattered fraction due to velocity detuning and Bragg scattering response function (Fig. 6.16). **I** Expected differential scattered atoms generated by multiplying **H** and Fig. 6.17E

dard error shown in Fig. 6.19B. As can be seen, the number of mislabelled vortices as a function of the standard error is monotonically increasing. This means that if the wrong vortex configuration is selected, it is more likely to have mislabelled one vortex than it is to have mislabelled many vortices due to the standard error increasing with the number of mislabelled vortices. This is important when working with experimental data, since noise is always an issue and can lead to errors.

One drawback of using an exhaustive search to optimize the metric is that the computation time doubles for every vortex that is added to the system, since it doubles the size of the search space. The typical runtime for 16 vortices is about 5 minutes which

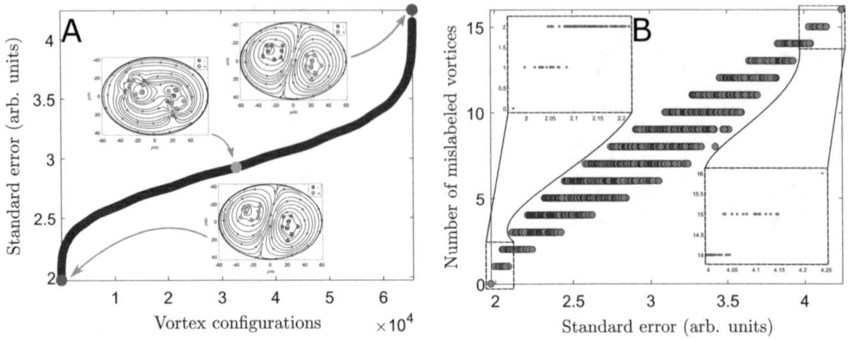

Fig. 6.19 Results from an exhaustive search of all possible 2^N vortex configuration for the vortex distribution and scattered atomic differential map shown in Fig. 6.17. **A** Shows the standard error for all possible vortex configurations with insets showing the vortex configurations that minimize, is the median, and maximize the standard error. **B** Number of mislabelled vortices versus the calculated standard error as can be seen the standard error scales almost monotonically with the number of mislabelled meaning the high number of mislabelled vortices are very unlikely

means that this method cannot be used in real time while running the experiment as it would quickly become intractable for more than 20 vortices. To get around this issue one can use the steepest-descent optimization algorithm presented in the next section.

6.5.6 Steepest-Descent Optimization

One of the realizations that led us to use steepest-descent as the method for optimizing the standard error is that we only care about the configuration that minimizes the metric and nothing else. Second is that to optimize the metric there is no need to limit the circulation of the vortices to be integer multiples of $\hbar/m$ since it is possible to generate differential scattered atoms maps (Fig. 6.18I) for non-integer vortex circulations. This leads to the idea that one could simply apply steepest-descent to find the circulation assignments that minimize the standard error. This is done by applying $\vec{\Gamma}_{n+1} = \vec{\Gamma}_n - \gamma_n \vec{\nabla} STE(\vec{\Gamma}_n)$ to each vortex n, where γ_n is the step size and $\vec{\nabla} STE(\vec{\Gamma}_n)$ is computed numerically, with the circulation bounded by $\pm\hbar/m$. The step size, γ_n, is allowed to change with every step using

$$\gamma_n = \frac{\left(\vec{\Gamma}_n - \vec{\Gamma}_{n-1}\right)^T \left[\vec{\nabla} STE(\vec{\Gamma}_n) - \vec{\nabla} STE(\vec{\Gamma}_{n-1})\right]}{||\vec{\nabla} STE(\vec{\Gamma}_n) - \vec{\nabla} STE(\vec{\Gamma}_{n-1})||^2} \tag{6.6}$$

with the initial step set to be a small size ($\gamma_0 = 0.01$). The initial vortex circulations guess is set to be 0. The descent is repeated until the following criterion is satisfied

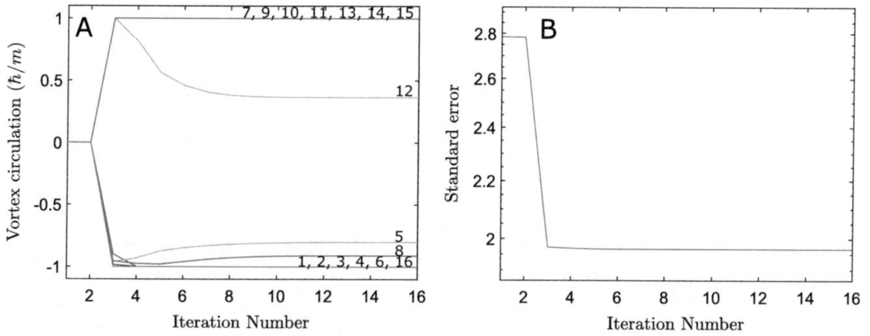

Fig. 6.20 Steepest-descent algorithm vortex circulation optimization. **A** Assigned vortex circulation for each vortex indicated (16 total), where the circulations are initially set to zero. **B** Standard error between the resulting differential map (Fig. 6.18G) and the measured one (Fig. 6.17F)

$$2\left|\frac{STE(\vec{\Gamma}_n) - STE(\vec{\Gamma}_{n-1})}{STE(\vec{\Gamma}_n) + STE(\vec{\Gamma}_{n-1})}\right| < \text{tolerance}, \tag{6.7}$$

where we usually set the tolerance to be 10^{-6}.

Figure 6.20A shows the assigned vortex circulation of all 16 vortices for every iteration of the steepest-descent when optimizing for the measured differential scattered atom map shown in Fig. 6.17F. It can be seen that most of the vortices converge to an integer circulation, but 3 of them have a final non-integer circulation. To get the vortex circulations that minimize the metric we simply assign each of the vortices with a circulation of $\pm\hbar/m$ based on the sign of their final optimized circulation. The fact that some vortices do not converge to integers simply means that we are not as certain of their signs as we are of the ones that have converged to the bound. Figure 6.20B shows the standard error versus the iteration number which is monotonically decreasing.

The big advantage of the steepest-descent optimization, as opposed to the exhaustive search of the space, is that the computation time is greatly reduced from 5 minutes to $\sim$1 s for the case of 16 vortices. This is mostly due to the fact that the steepest-descent computation time scales linearly with the number of vortices as opposed to exponentially. A potential drawback of the steepest-descent method is that there exists the possibility of converging to a local minimum instead of the global minimum, which would lead to the mislabelling of some vortices. Although we have not encountered this issue in our test cases, we cannot rule out this possibility.

6.5.7 Testing the Fractional Mapping

When generating the expected differential scattered atoms map, Fig. 6.18I, one of the main assumptions is that all of the detuning from the resonance comes from the

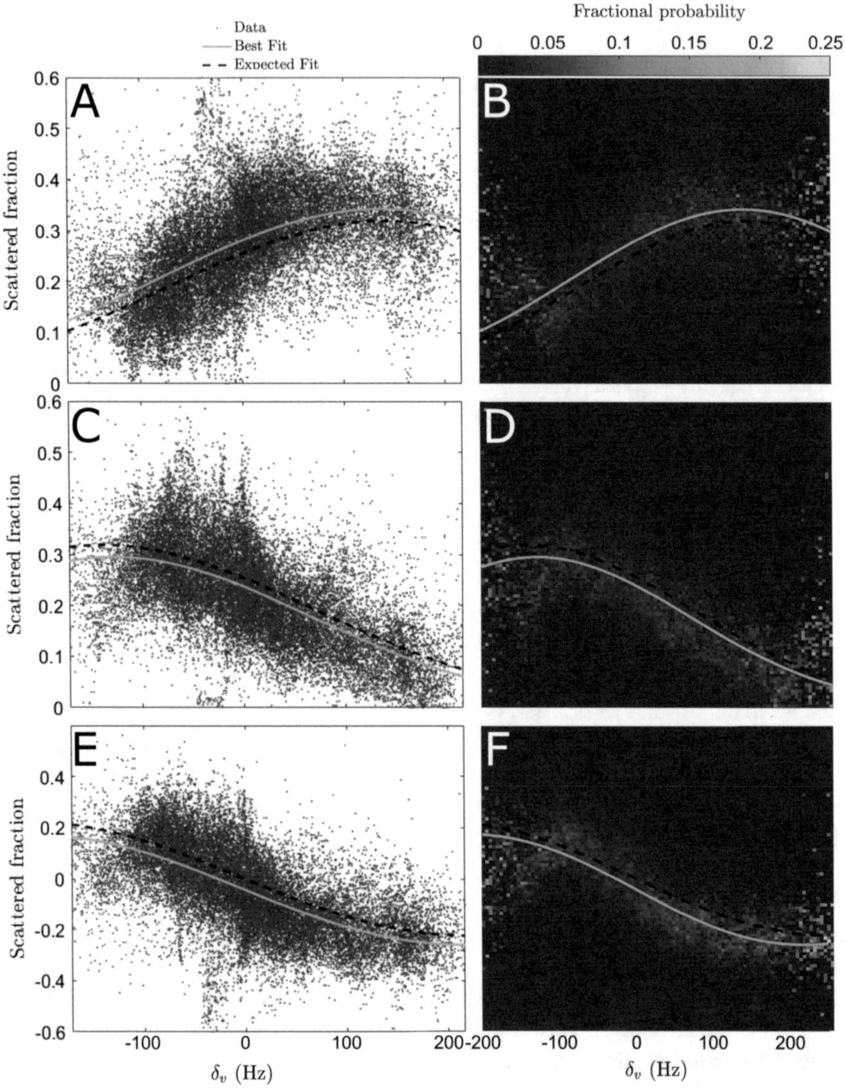

Fig. 6.21 Scattered measured double Bragg signal compared with the expected signal from Bragg scattered atom fraction measurement. **A** Plot of the top scattered atomic fraction Fig. 6.17B/E versus the Bragg frequency resonance detuning Fig. 6.18G. **B** Fractional probability of obtaining a certain scattered fraction given a Bragg frequency resonance detuning. **C**, **E** as in **A** but for the bottom Fig. 6.17D/E and differential Fig. 6.17F/E atomic fractions, respectively. **D**, **F** as in **B** but for **C**, **E**, respectively. The solid line represents the best fit to the data and the dashed line is the expected response from the measured Bragg scattering frequency response Fig. 6.16

velocity field imprinted by the vortices. This means that other processes that would shift resonance frequency, such as the stationary momentum distribution due to the finite extent of the cloud, and the mean field energy expansion along the Bragg beam propagation during TOF, have either been accounted for during the measurement of the Bragg scattering fractional frequency response [Eq. (6.5)] or are small compared to the quantities of interest. To verify that the mapping is approximately right, the measured atomic scattered fractions, and Bragg scattered atomic maps (Fig. 6.17B, D) are normalized to the pre-Bragg atomic map (Fig. 6.17E). These are plotted against the Bragg detuning, Fig. 6.18G, and are shown in Fig. 6.21A, C. The fractional probability of obtaining a given scattering fraction for a given detuning is shown generated from Fig. 6.21A, C are shown in Fig. 6.21B, D. The black dashed line represents the expected central probabilities using Fig. 6.16 and the red solid line represents the best fit to the data. As can be seen the Bragg fractional atom scattering frequency response Eq. (6.5) is a good approximate estimator of the most likely scattering fraction as a function of the Bragg detuning. The same exercise can be done in the case of the measured differential scattered atoms map and the results are shown in Fig. 6.21E, F where we use $S_{\text{frac}}(x, y) = \sigma(\delta_v(x, y) - \delta_0) - \sigma(\delta_v(x, y) + \delta_0)$, with $\delta_0 = 140$ Hz as the expected fit. The fit is accurate, confirming that Eq. (6.5) is a good estimator.

6.6 Conclusion

We have reported the first observation of the clustered phase of quantum vortices predicted by Onsager's statistical mechanics of point-vortices in a bounded domain [4]. Once achieved, the clustered phase is remarkably robust to dissipation, contrary to the conventional wisdom for negative temperature states. Meanwhile, the evaporative heating mechanism appears to be more fragile, inhibited by modest dissipation. Nonetheless, a systematic study of the clustering transition and its emergence from quantum turbulence [17–19] appears within reach, if further reduction of thermal dissipation can be achieved. The precise control of the trapping potential in our experiment enables a broad range of stirring and trapping configurations, opening the door to further studies of the vortex clustering phase transition [14, 17, 18], and of fully developed quantum turbulence confined to two dimensions. Emerging tools for precision characterization, including vortex circulation detection [24], momentum spectroscopy [41], and correlation functions [27, 42], can be expected to provide further insights into the role of coherent structures in 2D vortex matter.

References

1. Kraichnan RH, Montgomery D (1980) Two-dimensional turbulence. Rep Progress Phys 43:547
2. Montgomery D, Joyce G (1974) Statistical mechanics of "negative temperature" states. Phys Fluids 17:1139–1145
3. Kraichnan RH (1967) Inertial ranges in two-dimensional turbulence. Phys Fluids 10:1417–1423
4. Onsager L (1949) Statistical hydrodynamics. Il Nuovo Cimento 1943–1954(6):279–287
5. Lin CC (1941) On the motion of vortices in two dimensions. Proc Natl Acad Sci U S A 27:570–575
6. Maurer J, Tabeling P (1998) Local investigation of superfluid turbulence. EPL (Europhys Lett) 43:29
7. Henn E, Seman J, Roati G, Magalhães KMF, Bagnato VS (2009) Emergence of turbulence in an oscillating Bose-Einstein condensate. Phys Rev Lett 103:045301
8. Rutgers MA (1998) Forced 2D turbulence: experimental evidence of simultaneous inverse energy and forward enstrophy cascades. Phys Rev Lett 81:2244–2247
9. Smith RA, O'Neil TM (1990) Nonaxisymmetric thermal equilibria of a cylindrically bounded guiding-center plasma or discrete vortex system. Phys Fluids B: Plasma Phys 2:2961–2975
10. Binney J, Tremaine S (2011) Galactic dynamics. Princeton University Press
11. Young RM, Read PL (2017) Forward and inverse kinetic energy cascades in Jupiter's turbulent weather layer. Nat Phys 13:1135
12. Tabeling P (2002) Two-dimensional turbulence: a physicist approach. Phys Rep 362:1–62
13. Kosterlitz JM, Thouless DJ (1973) Ordering, metastability and phase transitions in twodimensional systems. J Phys C: Solid State Phys 6:1181
14. Yu X, Billam TP, Nian J, Reeves MT, Bradley AS (2016) Theory of the vortex-clustering transition in a confined two-dimensional quantum fluid. Phys Rev A 94:023602
15. Kraichnan RH (1975) Statistical dynamics of two-dimensional flow. J Fluid Mech 67:155–175
16. Fetter AL (1966) Vortices in an imperfect Bose gas. IV. Translational velocity. Phys Rev 151:100–104
17. Billam TP, Reeves MT, Anderson BP, Bradley AS (2014) Onsager-Kraichnan condensation in decaying two-dimensional quantum turbulence. Phys Rev Lett 112:145301
18. Simula T, Davis MJ, Helmerson K (2014) Emergence of order from turbulence in an isolated planar superfluid. Phys Rev Lett 113:165302
19. Salman H, Maestrini D (2016) Long-range ordering of topological excitations in a two-dimensional superfluid far from equilibrium. Phys Rev A 94:043642
20. Campbell L, O'Neil K (1991) Statistics of two-dimensional point vortices and high-energy vortex states. J Stat Phys 65:495–529
21. Neely TW, Samson EC, Bradley AS, Davis MJ, Anderson BP (2010) Observation of vortex dipoles in an oblate Bose-Einstein condensate. Phys Rev Lett 104:160401
22. Neely TW et al (2013) Characteristics of two-dimensional quantum turbulence in a compressible superfluid. Phys Rev Lett 111:235301
23. Kwon WJ, Kim JH, Seo SW, Shin Y (2016) Observation of von Kármán vortex street in an atomic superfluid gas. Phys Rev Lett 117:245301
24. Seo SW, Ko B, Kim JH, Shin Y (2017) Observation of vortex-antivortex pairing in decaying 2D turbulence of a superfluid gas. Sci Rep 7:4587
25. Groszek AJ, Simula TP, Paganin DM, Helmerson K (2016) Onsager vortex formation in Bose-Einstein condensates in two-dimensional power-law traps. Phys Rev A 93:043614
26. Rooney SJ, Blakie PB, Anderson BP, Bradley AS (2011) Suppression of Kelvon-induced decay of quantized vortices in oblate Bose-Einstein condensates. Phys Rev A 84:023637
27. White AC, Barenghi CF, Proukakis NP (2012) Creation and characterization of vortex clusters in atomic Bose-Einstein condensates. Phys Rev A 86:013635
28. Stagg GW, Parker NG, Barenghi CF (2014) Quantum analogues of classical wakes in Bose-Einstein condensates. J Phys B: At Mol Opt Phys 47:095304

29. Esler JG, Ashbee TL (2015) Universal statistics of point vortex turbulence. J Fluid Mech 779:275–308
30. Johnstone SP et al (2019) Evolution of large-scale flow from turbulence in a two-dimensional superfluid. Science 364:1267–1271
31. Moon G, Kwon WJ, Lee H, Shin Y-I (2015) Thermal friction on quantum vortices in a Bose-Einstein condensate. Phys Rev A 92:051601
32. Lin Y-J, Perry A, Compton R, Spielman I, Porto J (2009) Rapid production of $^{87\$}$Rb Bose-Einstein condensates in a combined magnetic and optical potential. Phys Rev A 79:063631
33. Rakonjac A et al (2016) Measuring the disorder of vortex lattices in a Bose-Einstein condensate. Phys Rev A 93:013607
34. Choi S, Morgan S, Burnett K (1998) Phenomenological damping in trapped atomic Bose-Einstein condensates. Phys Rev A 57:4057
35. Rooney SJ, Blakie PB, Bradley AS (2012) Stochastic projected Gross-Pitaevskii equation. Phys Rev A 86:053634
36. Gauthier G et al (2016) Direct imaging of a digital-micromirror device for configurable microscopic optical potentials. Optica 3:1136–1143
37. Dennis GR, Hope JJ, Johnsson MT (2013) XMDS2: fast, scalable simulation of coupled stochastic partial differential equations. Comp Phys Comm 184:201–208
38. Nore C, Abid M, Brachet ME (1997) Kolmogorov turbulence in low-temperature superflows. Phys Rev Lett 78:3896–3899
39. Esler JG, Ashbee TL, Mcdonald NR (2013) Statistical mechanics of a neutral point-vortex gas at low energy. Phys Rev E 88:012109
40. Powis AT, Sammut SJ, Simula TP (2014) Vortex gyroscope imaging of planar superfluids. Phys Rev Lett 113:165303
41. Reeves MT, Billam TP, Anderson BP, Bradley AS (2014) Signatures of coherent vortex structures in a disordered two-dimensional quantum fluid. Phys Rev A 89:053631
42. Skaugen A, Angheluta L (2016) Velocity statistics for nonuniform configurations of point vortices. Phys Rev E 93:042137

Chapter 7
Conclusion

In this thesis we have experimentally, numerically and theoretically studied the creation of versatile potentials for BEC condensate, along with studies of the fundamental properties of superfluid transport and 2D turbulence. In this chapter, we by briefly summarizing the results of Chaps. 4, 5, and 6 and conclude by presenting ongoing work in the lab while also suggesting future work that expands on the presented research.

7.1 Summary

Chapter 2 focused on the background knowledge needed to understand superfluid transport and turbulence in BECs while Chap. 3 described the apparatus used to perform the studies presented in this work.

Chapter 4 described expanded controls over superfluid BEC systems by using a directly imaged DMD device that projects light onto the plane of a highly oblate BEC. The technique was demonstrated to be suitable for implementation in both quantum gas experiments and other optical trapping applications. We showed examples of possible static potentials, and how halftoning and time-averaging could be used to produce potentials beyond simple binary patterns. We demonstrated the use of the high speed and precision of the DMD to create dynamic potentials, allowing for studies in time-dependent potentials. We presented a density-based feedback method to correct for aberrations of the projection system, both with a halftoning technique and a time-averaged one. Future applications and current studies using these non-binary potentials, such as vortex dipole optics, vortex trapping and current imprinting, were presented. DMD projection enabled our versatile apparatus to perform the studies presented in Chaps. 5, and 6.

© The Editor(s) (if applicable) and The Author(s), under exclusive license to Springer Nature Switzerland AG 2020

G. Guillaume, *Transport and Turbulence in Quasi-Uniform and Versatile Bose-Einstein Condensates*, Springer Theses,
https://doi.org/10.1007/978-3-030-54967-1_7

Chapter 5 investigates the dynamics of a dumbbell atomtronic circuit and demonstrates the similarity to acoustics, and how it can be modelled with lumped elements. This is done by first deriving an effective capacitance and inductance from the hydrodynamic equations and modelling the transport as a small perturbation on the ground state of the system, allowing for a Thomas-Fermi approximation. The low amplitude plasma oscillations were analysed, and it was found that modelling the inductance of the system requires careful consideration of the full flow field, although an analytic approach can be developed to assign the inductance for particular circuit elements. The frequency of the superfluid oscillation was also found to match the expected acoustic oscillation for the same system. The break down of superfluidity and its effect on the superfluid conductivity were studied, and an Ohmic conductive relationship across the entire parameter regime was found, suggesting that observing Feynman-like resistance will require a situation closer to this idealised model, such as remaining in the incompressible limit by maintaining quasi-uniform superfluid density.

Chapter 6 reports the first observation of the clustered phase of quantum vortices predicted by Onsager's statistical mechanics of point-vortices in a bounded domain. These clusters were generated using the sweep of a paddle-shaped obstacle through an oval-shaped hard-walled BEC without circular symmetry, leading to a preferred dipole moment along the major axis of the ellipse. The clusters were injected so as to minimize sound excitations and maximize their entropy, so the initial system would be near equilibrium. The signs of the vortices in the initial clusters were verified using momentum spectroscopy. In studying the dynamics of this clustered phase, we found it to be remarkably robust to dissipation, contrary to the conventional wisdom for negative temperature states. The dissipation mechanisms for the vortices were explored opening the door to the systematic study of the clustering transition and its emergence from quantum turbulence. These results are generally important for the study of fully developed quantum turbulence in two-dimensional superfluids.

7.2 Future Outlook

The versatility of the apparatus means that many future experiments can be performed. Here we list a few of the potential candidates.

Following from the quantum turbulence work presented in Chap. 6. Kwan Goddard Lee is currently looking at the symmetry breaking transition of a disk of containing N like-sign point voticies, whereby the cluster will be on-axis or off-axis depending on the angular momentum and energy [1]. The goal will be to observe nonaxisymmetric equilibria forming dynamically from a non-equilibrium initial condition. These initial non-equilibrium states are created by generating customized stirring obstacles using the DMD.

The utilization of the dynamic capabilities of the DMD to imprint deterministic currents through stirring a superfluid in a ring is being investigated. This technique is currently being contrasted with the use of a linear light gradient which can perform the same imprinting, but faster. The creation of a vortex box trap, presented in Chap. 4, presents an interesting geometry for superfluid turbulence experiments.

The apparatus has opened exciting doors for furthering our understanding of superfluids and their behaviour. It presents a nearly ideal system for the study of the 2D point vortex model, and should become better after the planned addition of an optical accordion as described in [2]. This update would increase our control over the energy cut-off in our trap, allowing for more deeply-cooled BECs with reduced thermal clouds. Thermal damping was found to be detrimental to the dynamics of superfluid transport in Chap. 5 and point-vortex dynamics in Chap. 6.

References

1. Smith RA, O'Neil TM (1990) Nonaxisymmetric thermal equilibria of a cylindrically bounded guiding-center plasma or discrete vortex system. Phys Fluids B: Plasma Phys 2:2961–2975
2. Ville JL et al (2017) Loading and compression of a single two-dimensional Bose gas in an optical accordion. Phys Rev A 95:013632

Appendix A
Tunable Atomtronic Circuit Data

In this section the data taken for the tunable atomtronic circuit study is tabulated. The system oscillation frequency is measured for a range of channel length, channel width and atom number. Table A.1 shows the frequency dependence on the length for a range of atom numbers, and Table A.2 shows the frequency dependence on width for a range of atom numbers. We extrapolate the frequency, to the fixed atom number N $= 2.00 \times 10^6$, using a fit to the atom number dependence of the form $f(N) = A \times N^B$, where A (B) are free parameters. The extrapolated frequencies are shown in the last row of the table.

For the resistive decay measurements, GPE and experimental channel density data for the simulated and measured channels is shown in Tables A.3 and A.4, respectively. The data is extrapolated from a fit to the population imbalance decay, starting with an initially full reservoir, of the form $\eta = A \exp[-t/\tau] + (1 - A)$, where A and τ are the fit parameters.

© The Editor(s) (if applicable) and The Author(s), under exclusive license to Springer Nature Switzerland AG 2020

G. Guillaume, *Transport and Turbulence in Quasi-Uniform and Versatile Bose-Einstein Condensates*, Springer Theses,
https://doi.org/10.1007/978-3-030-54967-1

Table A.1 Experimental data showing the oscillation frequency as a function of length for a channel of fixed width, 12.5 μm, for a variety of initial atom numbers. Bottom row indicates the frequency extrapolated to 2×10^6 atoms

Length (μm)					
1.5		3		6	
N ($\times 10^6$)	f (Hz)	N ($\times 10^6$)	f (Hz)	N ($\times 10^6$)	f (Hz)
0.79(4)	9.0(4)	0.73(4)	8.4(6)	0.89(4)	8.3(5)
1.08(4)	10.5(4)	1.25(6)	9.8(4)	1.16(3)	9.2(2)
1.60(8)	11.6(3)	1.54(9)	10.8(3)	1.67(4)	10.4(3)
1.76(9)	11.8(2)	1.79(4)	11.3(4)	1.85(8)	10.6(4)
1.94(8)	12.5(2)	–	–	–	–
2.00	12.4(7)	2.00	11.7(3)	2.00	10.9(5)
Length (μm)					
10		20		35	
N ($\times 10^6$)	f (Hz)	N ($\times 10^6$)	f (Hz)	N ($\times 10^6$)	f (Hz)
1.14(7)	8.8(5)	0.76(3)	6.7(4)	1.27(10)	6.0(4)
1.23(5)	6.5(4)	1.33(5)	7.8(5)	1.53(6)	6.6(4)
1.37(6)	9.2(2)	1.73(8)	9.0(2)	1.74(6)	6.9(4)
1.50(5)	9.4(4)	2.11(8)	9.2(2)	2.21(6)	7.2(2)
2.00	9.8(3)	2.00	9.1(3)	2.00	7.1(6)

Table A.2 Experimental data showing the oscillation frequency as a function of width for a channel of fixed length, 1.5 μm, for a variety of initial atom numbers

Width (μm)							
5		7.5		10		12.5	
N ($\times 10^6$)	f (Hz)	N ($\times 10^6$)	f (Hz)	N ($\times 10^6$)	f (Hz)	N ($\times 10^6$)	f (Hz)
1.00(5)	7.5(7)	0.81(9)	7.8(7)	0.40(2)	6.2(1)	0.79(4)	9.0(4)
1.53(4)	8.7(4)	1.50(5)	9.9(5)	0.81(6)	8.3(4)	1.08(4)	10.5(4)
2.51(2)	10.5(4)	1.60(3)	10.1(5)	1.02(5)	9.3(4)	1.60(8)	11.6(3)
–	–	2.06(5)	11.2(6)	1.87(2)	10.2(5)	1.76(9)	11.8(2)
–	–	–	–	–	–	1.94(8)	12.5(2)
2.00	9.7(6)	2.00	11.1(10)	2.00	11.0(10)	2.00	12.4(7)
Width (μm)							
17.5		22.5		30		40	
N ($\times 10^6$)	f (Hz)	N ($\times 10^6$)	f (Hz)	N ($\times 10^6$)	f (Hz)	N ($\times 10^6$)	f (Hz)
0.46(2)	8.3(6)	0.81(6)	11.0(11)	1.09(6)	11.0(10)	1.24(7)	14.5(13)
1.46(4)	12.3(5)	1.80(5)	12.6(8)	1.82(8)	13.8(10)	1.72(5)	16.9(7)
1.63(4)	12.6(5)	1.96(8)	13.4(14)	2.05(10)	14.2(4)	2.07(6)	17.3(7)
2.00	13.7(7)	2.00	13.3(3)	2.00	14.8(8)	2.00	17.2(8)

Table A.3 GPE resistive transport parameters for a fixed 10 μm length channel

Width (μm)	τ (s)	C ($\times 10^{35}$ J^{-1})	n_{1D} (μm^{-1})	ξ (μm)
2	1.517	5.98	83	0.581
2.5	0.644	5.99	175	0.443
3	0.346	6.00	287	0.376
3.5	0.249	6.01	384	0.370
4	0.182	6.02	553	0.342
4.5	0.150	6.03	687	0.321
5	0.119	6.03	844	0.291
6.25	0.085	6.05	1.11×10^3	0.284
7.5	0.061	6.07	1.48×10^3	0.278
8.75	0.052	6.09	1.78×10^3	0.280
10	0.044	6.11	2.09×10^3	0.275
12.5	0.033	6.15	2.75×10^3	0.270
15	0.027	6.19	3.37×10^3	0.269

Table A.4 Experimental resistive transport parameters for a fixed 10 μm length channel

Width (μm)	τ (s)	C ($\times 10^{35}$ J^{-1})	n_{1D} (μm^{-1})
0.75	4.535	5.97	40
1	2.624	5.97	84
1.5	0.840	5.98	130
2	0.280	5.99	454
2.5	0.213	5.99	643
5	0.101	6.03	990
7.5	0.049	6.07	2.000×10^3
10	0.030	6.11	2.800×10^3
15	0.020	6.19	4.312×10^3

Printed in the United States
by Baker & Taylor Publisher Services